Success in Farming

Dr. M. Upton

Success in Farming

M. Upton

Donated to the University of Reading Library by the author.

Martin Upton

October 1994

MACMILLAN PUBLISHERS

First published 1985

Published by *Macmillan Publishers Ltd*
London and Basingstoke
Associated companies and representatives in Accra, Auckland, Delhi, Dublin, Gaborone, Hamburg, Harare, Hong Kong, Kuala Lumpur, Lagos, Manzini, Melbourne, Mexico City, Nairobi, New York, Singapore, Tokyo

ISBN 0 – 333 – 38335 – 4

Printed in Hong Kong

P.
Professor M. Upton.

Contents

1 Factors of Production

1.1 The Farm as a System

We all depend upon the farmer for our food and raw materials for industry. In many African countries more than half the population are employed in farming. Yet it is generally found that farmers' incomes, on average, are below urban wage levels and the gap may be increasing. Many young people leave agriculture and go to the towns in search of jobs. There are not enough jobs for all these migrants and many finish up unemployed. At the same time, slow growth of agricultural production may result in food shortages and rising food prices. To avoid these problems it is important that agricultural production and agricultural income grow at least as fast as the rest of the economy. Success in economic development depends upon success in farming.

A farm is an area of land where a farmer, farm family or other organization, cultivates crops or tends livestock. The size may range from a small subsistence plot of less than one hectare to a commune or state farm comprising all the land of several villages. The pattern of farming may be based on shifting cultivation, fallow farming or permanent cultivation. It may involve irrigation or perennial tree crop plantations. It may include grazing cattle, sheep or goats or keeping livestock intensively such as pigs or poultry. All these are different farming systems.

It is useful to think of a farm as a system in which inputs such as seed, soil, water, fertilizer and human effort are converted into outputs such as grain. (See Fig 1.1)

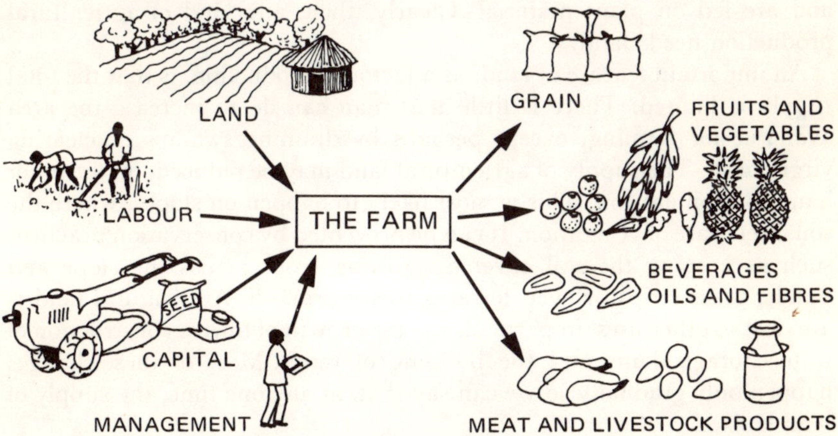

Fig 1.1 The farm as a system

For livestock, inputs of animals, food, water, medicines and human effort are converted into milk, or meat. Most farming systems consist of several different lines of production, or 'enterprises' which are interlinked. For instance under 'mixed farming' both crop and livestock enterprises are combined. Crop residues, which are outputs of the crop enterprise, are fed to livestock, and therefore become inputs to that enterprise. Manure from the livestock may then be spread on the land as an input to the crop enterprise. Because of many such linkages, a farming system is quite complicated, and a small change to one enterprise may cause other changes throughout the system.

The kind of output which is produced varies from one farm to another, or at least from one climatic zone to another. Rice or root crops may be produced in the wetter tropics, while cereals or groundnuts may be produced in drier zones. Cattle production too may be concentrated in the more arid areas. However, there are four main sources of inputs that are found on every farm in every part of the world. These are land, labour, capital and management, often called the four 'factors of production'. They are all needed for agricultural production to take place and each has its own special features, which we will consider in more detail.

1.2 Land

The growth of crop plants depends upon a supply of air, light and warmth from the sun, rainwater and soil minerals. All these are free gifts of nature, available along with the land on which the crops are grown. Similarly the wild grasses and browse of grazing range lands are 'natural resources', which man exploits for cattle, sheep and goat production. Even intensive livestock, such as pigs or poultry, need a certain amount of space for living and are fed on plant material. Clearly, then, any kind of agricultural production needs land.

An important feature of land, as a factor of production, is that the total supply is limited. There is little that man can do to increase the area available for farming, except perhaps by draining swamps or clearing virgin bush. The supply of agricultural land may be reduced, on the other hand, by soil erosion. This is most likely to happen on slopes, where the soil is left bare of vegetation. It can be prevented by conservation practices such as keeping the soil covered, growing crops across the slope and bunds, grass strips or even terraces to stop run-off. Agricultural land is also lost to other uses, in particular to the growth of towns, construction of water-storage dams and the building of roads. Most of these changes happen only gradually so we can say that, at any one time, the supply of land is fixed.

The amount of land each individual farmer or farm family can acquire is determined by the system of land tenure. This is the set of customs or laws by which land rights and duties are allocated. Farm size must also be influenced by the density of the farming population. Under traditional, customary tenure every adult citizen is entitled to a piece of land in his home village for the cultivation of crops. He also has the right to graze livestock, hunt and gather firewood from communal land. This seems a just and fair system of allocation but it has some disadvantages. Some villagers who move to the town may jealously keep their rights even though they no longer farm the land. If the rights are inherited and split up among family members, a farmer may be left with several small scattered holdings which are inconvenient to farm.

As land becomes more valuable, either because the population becomes more dense or because profitable crops may be grown, a market for land develops. Land may be,

(i) bought or sold under freehold tenure
(ii) rented under a cash tenancy or leasehold
(iii) pledged as security for an interest free loan.

For any of these systems the amount of money paid per hectare is likely to depend on the returns that can be earned. Very fertile land is likely to cost more to buy or rent than poor land. Also the size of loan that can be secured by pledging a hectare of rich land, will be bigger.

Land is also immobile. It cannot be moved to a more convenient position. This means that land near the village centre or even near a road, is more valuable than inaccessible plots. Less time is spent taking inputs to the fields or carrying produce to the market. As a result, higher returns can be earned. Similarly, river valley land conveniently placed for irrigation may be valued and priced more highly than upland.

Consideration of land and natural resources reminds us that the farmer is very dependant on the whims of nature. If the rains come late or fail later in the season, his crops might fail. A crop or livestock disease outbreak, or an insect pest can cause serious losses. Farming systems are complicated by the need to avoid risk or at least to limit its impact.

1.3 Labour

Farming is a human activity. Work is involved in clearing land, caring for crops and livestock and transporting produce to market. Even where most of the tasks are mechanized, human drivers and machine operators are needed. On many small farms the labour force consists only of family members; the farmer, his wife and children. Neighbouring families may help each other from time-to-time when there is heavy work to be done such as bush clearing. This is really an exchange of labour, since the help is always returned at some later stage. Although it may be done more quickly and easily by a large gang, the exchange of labour does not increase the total supply. On larger farms labour is hired, either on a regular, full-time basis or on a temporary or part-time basis. People, whether they are family members or hired workers, will work harder and more productively if they are healthy, contented and well motivated. So a farmer should ensure that his workers have adequate diets, avoid debilitating diseases and are respected as individuals and made to feel part of a team. Good staff management is needed for success in farming.

Within the family or the hired work force there may be scope for 'division of labour'. This means that different persons specialise in different tasks according to their skills and abilities. By specialising and repeating certain tasks they become even more skillful and learn to work faster and better. Within the family, men are usually responsible for heavy tasks such as bush clearing and soil cultivation while women and children do the weeding, harvesting and processing at which they may become more skilled than the men. When labour is hired, some are hired as field workers, others as tractor drivers and others as stockmen.

This division of labour makes the measurement of labour inputs difficult. The unit which is commonly used is the man-day (the work of a man for an average working day) or the man-hour. But, clearly a man-day of tractor driver's time is different from a man-day of stockman's time. Similarly, the amount of work done by a woman or a child in an hour may differ from the work done by a man. Therefore when planning a farm system it may be best to think about the work input for each category of labour separately. For convenience, in this book, we will use the man-day to mean a day's work by any person, and similarly with the man-hour.

Because of the seasonal variation in rainfall and crop growth, the farm work load varies from season to season. For example, the seasonal variation in man-hours worked per month on one farm is shown in Fig 1.2. The busiest period seems to be early in the year, about the start of the rains, when land clearing, cultivation and planting take place. On this farm more temporary labour is hired during the work peak than at other times of the year. This helps to even out the work load for the family. Nonetheless, it is usually difficult to keep family members or regular, full-time hired workers fully employed throughout the year. Labour is only scarce during the work peaks; at other times there is seasonal unemployment. The seasonal variation is small where continuous, sequential cropping is possible under irrigation for instance. A mixture of enterprises generally has a more even seasonal work load than a single crop does.

Hired workers are paid wages either in cash or in kind. It should therefore be possible to work out the daily or hourly cost of hired labour.

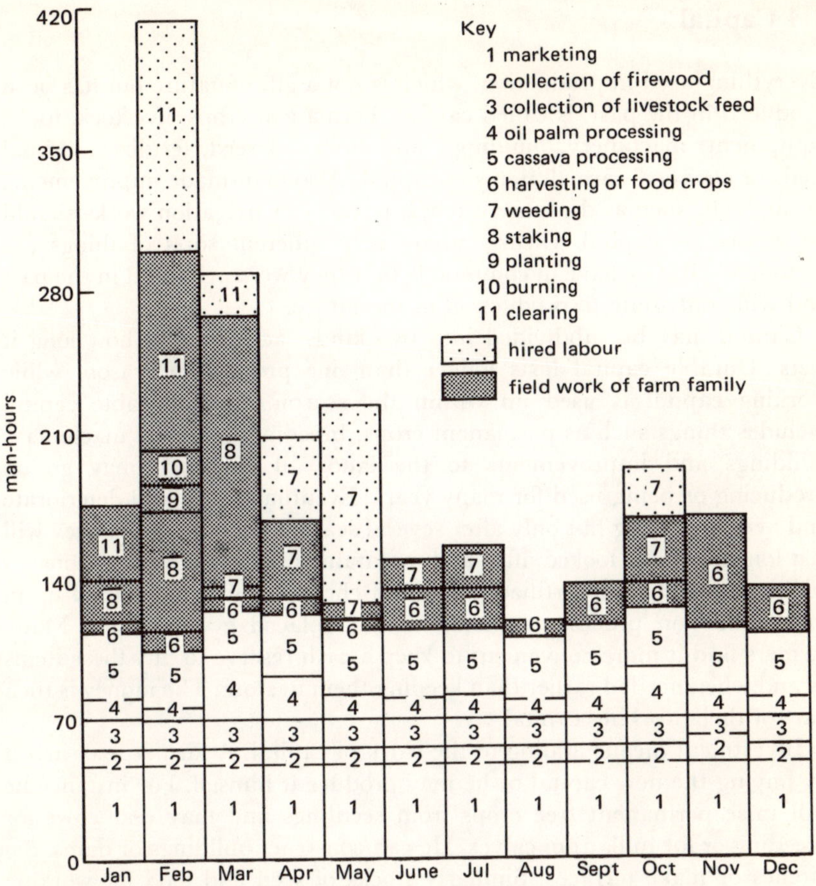

Source: Lagemann, J., Flinn, J.C., Okigbo, B.N. and Moorman, F.R.: op.cit., p.23.

Fig 1.2 Distribution of farm activities over one year

This can be compared with the returns to decide whether it is worth hiring. Family labour also has a cost. For a start everybody needs food and the basic necessities of life to survive. Also family members have many other ways of spending their time. First there are many daily household tasks to do such as cooking and cleaning, fetching water and firewood. Children may have to attend school. Then there are other activities which occur occasionally such as housebuilding, furniture making, attending ceremonies and festivals. Finally there may be other profitable ways of spending time in trading or some other off-farm job. So it may be argued that time spent on the farm has a cost, which is the value of what could be earned in an alternative, off-farm activity. This is known as the 'opportunity cost' of labour. It is the cost in terms of alternative opportunities given up.

1.4 Capital

Everything used in production, which is not a gift of nature but has been produced in the past, is called capital. Permanent crops, livestock, tools, equipment, machinery, buildings, and stocks of seed, fertilizer, animal feeds or chemicals are all items of capital. Also man-made improvements in the land, such as drainage ditches, terraces or irrigation works should be viewed as capital. Clearly many very different sorts of things are included. All they have in common is that they were produced in the past and will contribute to production in the future.

Capital may be subdivided into two kinds, according to how long it lasts. Durable capital lasts longer than one production season, while working capital is used up within the season. Thus durable capital includes things such as permanent crops, breeding livestock, machinery, buildings and improvements to the land, all of which may go on producing or being used for many years. Eventually they will deteriorate and need replacing but only after several seasons use. Generally they will last longer, if well looked after and maintained. Working capital consists of the stocks of seed, fertilizer, food and chemicals which are used up in the production process and need to be replaced every season. Many farmers find it more convenient to keep a cash reserve to buy these items as and when needed rather than keeping them in store. The money is then part of their working capital.

Investment means adding to the stock of capital. A farmer may invest by buying the new capital or he may produce it himself. For instance he will raise permanent tree crops from seedlings and may rear cows for breeding or for milk from calves. He can construct buildings or dams, dig ditches or make terraces. Similarly stocks of seed and food for working capital may be kept from the previous harvest. To buy new capital, he must either draw on his savings or borrow the money.

In fact all capital investment involves a cost. Even if the capital is produced on the farm, there is a cost in terms of the work involved and perhaps other things the family have had to do without. There is an opportunity cost of making the investment. Having met this cost the farmer then has to wait for the benefits. Benefits may come in terms of a bigger output in the future or lower costs, for instance if labour is saved. The advantage is that the future benefits, when they do come, add up to more than the investment costs. In other words, the farmer gains by using roundabout methods of production. The net gain, after meeting all the current costs and making some allowance for the original investment cost is known as the **return on capital** or the **return on investment**.

Capital is made up of so many widely different items, the only convenient way of measuring the amount employed or the return on investment, is by using money values. This raises problems, because for the items produced on the farm, such as drainage ditches, or growing tree crops there is no market price on which to base the valuation. Items which are purchased on the other hand start to lose value immediately. A second-hand tractor is worth less than a new one. This loss in value is known as **depreciation.** Unless there is a well developed trade in second-hand capital items, it may be difficult to estimate their value or the rate of depreciation.

Actually the cost of using a piece of capital, such as a tractor, includes three elements. These are:

(i) the operating costs, fuel, oil and drivers' wages,
(ii) the cost of the tractor itself which may be averaged over its lifetime to estimate the annual depreciation,
(iii) the cost of having money tied-up in the tractor, which on borrowed money is the interest paid. Alternatively it may be possible to hire tractor services or those of some other kinds of capital. The hourly hire charge covers all the cost elements (unless it is subsidised) and is generally well known to farmers.

Despite the difficulties of measuring quantities of capital it is fairly obvious that more capital is used on commercial, mechanized farms than on traditional small holdings. The capital of a subsistence cultivator may consist only of a few hoes, cutlasses, axes and baskets. In fact the introduction of new technology, in the form of seeds of new high yielding varieties, improved breeds of livestock, agricultural chemicals or new types of equipment, involves capital investment. Farmers may not have sufficient savings to make the investment. Therefore the provision of loans through agricultural credit schemes may be necessary to allow the spread of new technology and the growth of agricultural output.

1.5 Management

Try to imagine a farm, with land and capital resources and a farm staff ready and able to do the work, but with nobody willing to make decisions and organise the work. No production will take place, even worse perhaps crops and animals will die because they are not cared for. Management is another essential factor of production. No matter how small or large the farm is, someone has to provide a management input. The task may be shared, of course: On a family farm, wives and children may manage their own plots or particular crops. On a commercial farm responsibility may be shared between sectional managers or between the general manager and stockmen or field foremen. Ideally responsibility for decision making should be allocated to make best use of the available management skills. Success in farming depends on good management.

Good management involves the following tasks:

(i) Setting objectives That is deciding on goals or targets to be achieved. If a manager is hired, the objectives may be laid down for him by the owners. As years go by, and the manager gains experience his views are likely to be taken into account in modifying the objectives.

(ii) Decision making Thinking ahead, making plans and day-to-day decisions on what needs to be done.

(iii) Implementation Seeing that the plans and decisions are carried out.

(iv) Control Keeping records and accounts, and checking on whether objectives are being met. If problems are discovered quickly then solutions can more readily be found.

These four main management tasks are not just done once but must be carried on all the time in the light of changing circumstances. Flexibility and readiness to change is also important for good management. On a commercial farm there are four main areas where management is needed. These are production, marketing, financial and staff management. More will be said about the first three of these in later chapters of this book. Good staff management depends on motivating workers to do their best. Good wages and working conditions can help but personal relationships are important. More can be learned about human relationships from experience than from books.

As we have seen, the first task of management is setting objectives. Before we can think about how to achieve success in farming, we need to know what the farmer's objectives are. How does he measure success? Probably the first and most basic objective for the farmer, or for anyone else, is to meet food and other essential needs of his family. This is the most important goal for a small subsistence farmer. Once a man is fairly confident he can meet these needs without fail, other objectives may become more important. Most people have as another goal the respect and friendship of their neighbours. Some even seek for power and prestige through local politics or in other ways. Most farmers also aim to make profits. They want to be sure that their income from sales or outputs is bigger than the cost of the inputs used. Indeed their objective may be to make this difference as big as possible; to maximise profit. Money can, of course, be used to buy food and other necessities as well as power and prestige, so profit maximising may in fact help to achieve other objectives. Profit maximization may not be the only objective but for the commercial farmer it is important. For the rest of this book we will assume the main aim is to make profit.

2 Basic Economic Principles

2.1 The Law of Diminishing Returns

Economics is concerned with choices in the allocation of scarce factors of production between alternative uses; to satisfy human wants. It applies to choices at all levels from national and local government policy making down to individual household decisions. The basic principles of economics are therefore relevant to farm management decision making.

One important basic principle is the law of **deminishing returns** which really emphasises the problem of scarcity. Because some factors of production are limited in supply, there are limits on the amount of output that can be produced. This may seem to be stating the obvious, but it is worth using an example to show what this implies.

Let us consider the effect on maize yield of varying the level of fertilizer application on a fixed area of land. We assume that a farmer has carried out a small trial. He has marked out four plots of land, identical in size and in every other respect and planted maize on them all. However whereas he used no fertilizer on plot A, on plot B he applied one bag, on plot C he applied two and on plot D he applied three bags of fertilizer. At harvest he weighed the yield from each plot, to find out the effect of varying fertilizer use. Possible results of such a trial are shown in Table 2.1 and in Fig 2.1.

Table 2.1 The effect of fertilizer use on maize yield

	Plot yield (bags of maize)	Increase in yield (bags of maize)
Plot A (no fertilizer)	4	–
Plot C (two bags of fertilizer)	7	3
Plot B (one bag of fertilizer)	9	2
Plot D (three bags of fertilizer	9½	½

The first column of the table shows the yield per plot which increases as more fertilizer is applied. The second column shows the amount by which the yield increases for each additional bag of fertilizer. This is known as the **marginal product** per bag of fertilizer.

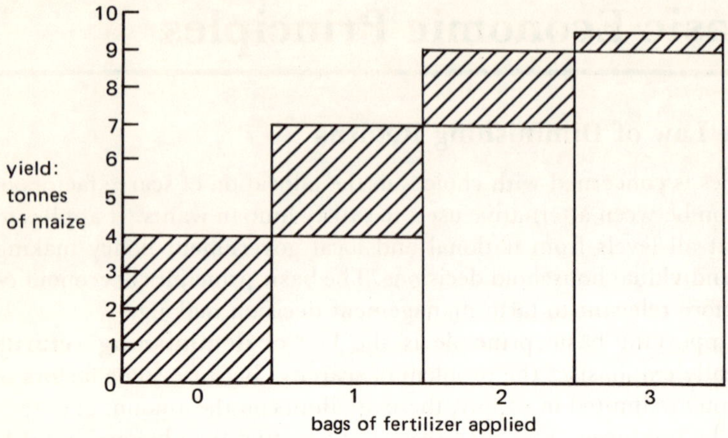

Fig 2.1 Diminishing returns

In Fig 2.1 the shaded blocks represent the marginal product for each additional bag of fertilizer. This example shows that the marginal product gets smaller (diminishes) as the amount of fertilizer is increased. It illustrates the law of diminishing returns which we can now state precisely. **If one input (fertilizer) is varied, the amounts of all other inputs (land, maize seed, labour) being held fixed, the marginal product per unit of the variable input (eventually) diminishes.** The word eventually should be included, because to start with, at low levels of variable input the marginal product may be constant or may even increase. But with increasing use of the variable input, beyond some point the marginal product must diminish.

This example illustrates the problem of scarcity. Because the amount of land, seed and labour are fixed, there is a limit to the amount of extra maize that can be produced, by increasing fertilizer use. The third bag of fertilizer only adds half a bag to the total yield of maize. It is quite likely that a fourth bag of fertilizer would not increase the yield at all. It might even be depressed. So the supply of maize is limited.

We may assume that this law applies whenever some inputs are varied while others are fixed. For instance, it may apply when the farming population increases on a fixed area of land. As the labour input rises the marginal product per person is likely to fall, unless new more productive systems of farming can be found.

2.2 The Economic Optimum

That is not the end of the story. If we know the effect of increasing fertilizer use on maize yields, and if we know the prices of fertilizer and maize, we can work out the most profitable rate of fertilizer application. This is known as the **economic optimum.**

To illustrate this let us assume that the price of fertilizer is $10 per bag and the price of maize is $4 per bag. At these prices it is only worth applying one bag of fertilizer. The marginal value product or extra income received from using one more bag of fertilizer is the marginal product times the price of maize. For the first bag of fertilizer this is 3 × $4 = $12. This is more than the cost of the fertilizer ($10) so it is profitable to use the first bag of fertilizer. The marginal value product of the second bag of fertilizer is 2 × $4 = $8 which is less than the cost of the fertilizer so it is **not** profitable to use the second bag. Clearly, it is only profitable to increase fertilizer use so long as the marginal value product is bigger than the cost per bag.

Actually the farmer does not necessarily have to use a whole number of bags of fertilizer. This means he can adjust the amount used more exactly, so as to maximise profit. Suppose he could vary the amount used by a tenth of a bag at a time. The cost of each unit is then $1. He should then use more fertilizer so long as the marginal value product is more than $1. The exact profit maximising position is found where the marginal value product is equal to the cost per unit. In our example this might be found where 1½ bags of fertilizer are used and a yield of 8 bags of maize is obtained. This is the economic optimum rate of fertilizer use. It should be noted that the optimum, or best, choice of fertilizer use is well below that needed for maximum yield. Profits cannot be maximised by maximising yields.

Now let us consider the effect of an increase in maize price to $5 per bag, while fertilizer still costs $10 per bag. The marginal value product of the first bag of fertilizer is raised to 3 x $6 = $18, while that of the second becomes 2 × $6 = $12. Since this is greater than the cost of $10, profit is maximised by applying at least two bags of fertilizer. The economic optimum output of maize is raised from 8 bags to over 9 bags. This illustrates a general principle that if the product price rises in relation to input costs, it is profitable to increase the product output. Product supply is increased.

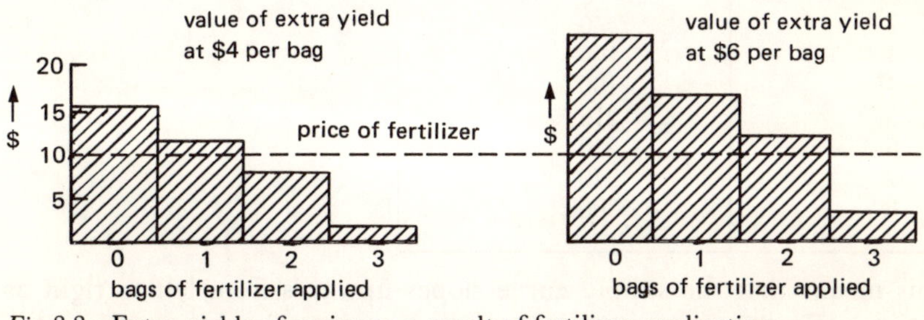

Fig 2.2 Extra yields of maize as a result of fertilizer application

Figure 2.2 may help to explain the effect. This shows the extra cash earned from each bag of fertilizer, first when maize is priced at $4 per bag, then when maize is priced at $6 per bag.

What happens on the other hand if the price of fertilizer rises? Consider the situation when maize sells for $4 per bag and fertilizer costs $15 per bag. Now it is not even worth applying one bag of fertilizer since the marginal value product is only $12. It might still be profitable to apply a fraction of one bag, but the main point to note is that the rise in input price has caused a reduction in the amount a farmer would wish to use. When input price rises the quantity used falls.

2.3 Market Supply

The word **market** has two rather different meanings. One meaning is a place where people meet at regular times to buy and sell. The other meaning refers to the total sales and purchases of a particular item in a region, the whole country, or even the whole world. So *the Nigerian maize market* means the total sales and purchases of maize throughout Nigeria over a given period. This second meaning is used here.

Market supply means the amount of the item that is offered for sale at a particular price and over a particular period. If market price rises the amount supplied normally increases. Let us stay with the example of maize supply. We have already seen that, for an individual farmer, a rise in maize price makes it profitable to increase maize output. In addition to this a rise in price may encourge other farmers to start growing maize in place of other crops. Furthermore, farmers will be encouraged to sell maize rather than eating it at home. For all these reasons the quantity supplied increases with increasing price, as shown in Table 2.2.

Table 2.2 Market supply schedule for maize

Price ($ per bag)	Quantity that would be supplied per year ('000 bags)
2	0
3	100
4	180
5	216
6	250
7	270
8	300

This means that the supply curve slopes upwards from left to right as shown in Fig 2.3

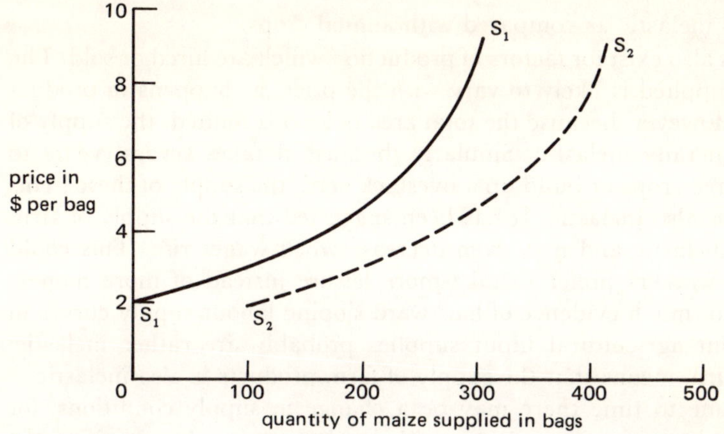

Fig 2.3 Supply curves for maize

The quantities shown in Table 2.2 and Fig 2.3 are imaginary, but national statistics usually show similar effects. It is important for government policy makers to know how big is the effect of changing prices on the amount supplied. Economists have a measure for this and it is called **elasticity.** If a small rise in price causes a big rise in quantity supplied, we say the supply is **elastic.** If, on the other hand, even a large increase in price has little effect on supply we say it is **inelastic**. More precisely the price elasticity of supply is defined as:

$$\frac{\text{Percentage change in quantity supplied}}{\text{Percentage change in price.}}$$

Supply is relatively elastic if the percentage change in quantity is bigger than the percentage change in price; that is if the elasticity is greater than one. Supply is relatively inelastic if elasticity is less than one. From the information given inTable 2.2 we can see that if price increases from $4 to $5 (by 25 per cent) the supply increases from 180 to 216 thousand bags (by 20 per cent). The elasticity is therefore 20 divided by 25 = 0.8. Supply is inelastic. This can be seen from the supply curve Fig 2.3 because the slope is increasingly steep. Roughly speaking a steep curve suggests inelastic supply and a flat curve suggests elastic supply.

The elasticity of supply varies according to the time period being considered and from one product to another. For instance over a very short period of say a week, the supply of maize is highly inelastic. The amount available is limited to the quantity in store at the time. Over a longer period, of a year for example, farmers have sufficient time to vary the area planted so the supply is more elastic. Clearly it is less easy to vary the area planted to a tree crop such as cocoa. This means that the supply of cocoa is inelastic as compared with annual crops.

Markets also exist for factors of production which are hired or sold. The quantity supplied is likely to vary with the price, as happens in product markets. However, because the total area of land is limited, the supply of land is generally inelastic. Similarly, because it takes several years to establish tree crops or build up a livestock herd, the supply of these items of capital is also inelastic. It has been suggested that the supply of farm labour is inelastic and may even decrease when wages rise. This could happen if workers prefer to take more leisure instead of more money. There is not much evidence of backward sloping labour supply curves in practice but agricultural input supplies probably are rather inelastic. This, in turn, means that the supply of farm products is also inelastic.

From time to time there may be a change in supply conditions, for instance if there is a particularly good growing season. At any price, the quantity supplied increases so the whole supply curve shifts to the right. In Fig 2.3 an increase in the supply of maize is shown by the movement of the curve from S_1 S_1 to S_2 S_2. Apart from these seasonal changes in supply there are other causes of shifts in supply of farm products. These include:

(i) Chance Poor rains, pests or diseases can cause supply to fall,

(ii) Changes in price of other items A fall in the price of fertilizer or of an alternative crop might cause farmers to increase maize production,

(iii) New suppliers joining the market Produce from a new government scheme or the growth of imports could raise supply, (note that unless there are import controls, supply of imports is usually very elastic),

(iv) Technological change Improved varieties or new methods of production can also lead to increases in supply.

Any of these shifts can cause changes in the prices farmers receive for their produce. But prices also depend on the market demand.

2.4 Market Demand

Market demand means the amount of an item that people are willing to buy at a particular price and over a particular period. If market price rises the amount demanded normally falls. For instance, if the price of maize rises people will buy less, partly because they cannot afford to buy as much as before and partly because they will switch to buying other cheaper foods instead of maize. Conversely if the price of maize should fall, people will buy more. For example, Table 2.3 shows quantities of maize demanded at different prices. These numbers can be used to draw a demand curve as in Fig 2.4. The supply curves are also shown here but should be ignored for the time being.

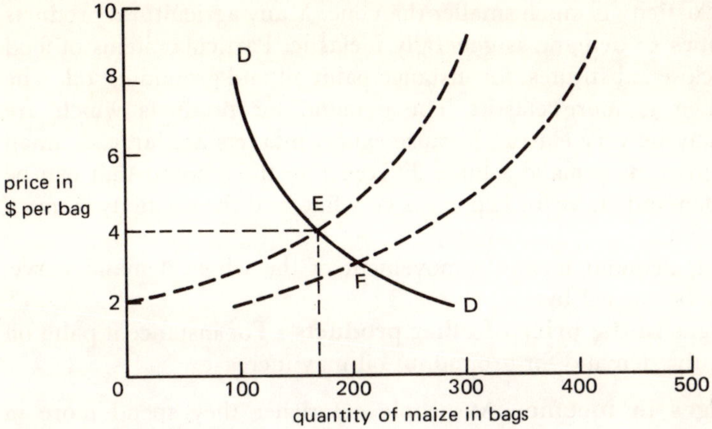

Fig 2.4 Demand curves for maize

Table 2.3 Market demand schedule for maize

Price ($ per bag)	Quantity that would be purchased per year ('000 bags)
2	280
3	220
4	180
5	150
6	125
7	110
8	100

Again it is very useful to know whether the demand is elastic in response to a price change. The price elasticity of demand is defined as:

$$\frac{\text{percentage change in quantity demanded}}{\text{percentage change in price}}.$$

Now price and quantity move in opposite directions, we must think of the elasticity as the fall in quantity demanded when price rises, or the rise in quantity when price falls. For example, Table 2.3 shows that when price falls from \$4 to \$3 (by 25 per cent) the quantity demanded rises from 180 to 220 thousand bags (by 22.2 per cent). Elasticity of demand is therefore $22.2 \div 25 = 0.89$. This is slightly below one so demand is slightly inelastic.

Elasticity of demand varies from one product to another. If it is a necessity, such as food without which people cannot survive and if there are few substitutes, demand is inelastic. The demand curve has a steep slope and elasticity is much smaller than one. Many agricultural products are necessities so demand is generally inelastic. Particular items of food may have close substitutes, for instance palm oil and groundnut oil. The demand then is more elastic. The demand for products which are exported may be very elastic, because export markets are large. A small change in price may make a huge difference to the amount that can be sold. The demand curve for exports is very flat, and the elasticity is much larger than one.

A shift in demand means a movement of the whole demand curve, which may be caused by;

(i) Changes in the price of other products For instance if palm oil price rises, the demand for groundnut oil may increase.

(ii) Changes in income As people get richer they spend more in general although spending on food may not increase as much as on other items.

(iii) Changes in population A growing population means more mouths to feed.

(iv) Changes in taste Advertising a product may persuade people to buy more.

In markets for factors of production, the demand comes from producers. We saw, near the beginning of this chapter, that if the price of fertilizer rises, farmers are likely to use less. The same is true of land, labour and all items of capital. Elasticity of demand for factors of production depends partly on the demand for the product. If the demand for food is relatively inelastic, the demand for labour used in producing food will be inelastic too. But factor substitution is also important. If machinery can be substituted for labour, then when wages rise farmers may switch to mechanized methods. This will make demand for labour more elastic.

2.5 Market Prices

The equilibrium market price of any product is the price level where quantity supplied is just equal to quantity demanded. This quantity is known as the **equilibrium quantity**. For the maize market example, given in Table 2.2 and 2.3, the equilibrium price is $4 per bag. At this price the supply of 180 000 bags is exactly the same as the quantity demanded, so this is the equilibrium quantity. In Fig 2.3 the equilibrium price and quantity are found at point E, where the supply and demand curves cross. At this point supply and demand just balance, so we normally expect the actual price paid to be near the equilibrium.

To see why, imagine what would happen in the maize market at a price of $3 per bag. Only 100 000 bags would be supplied while people would want to buy 220 000 bags. Buyers would start bidding the price up, and sellers would find they could charge more than $3 per bag. This would attract more sellers and put some buyers off until the equilibrium price and quantity was reached. Sometimes, governments set price controls (perhaps to keep food prices low) but this can cause problems. For instance, in this case, suppose the Government set the price of maize at $3. There would be an excess demand of 120 000 bags (220 000 – 100 000 bags). Some way of rationing out the 100 000 bags supplied would be needed. Also an illegal, black market might grow, with prices higher than the official price.

On the other hand, if the price rises above the equilibrium price there is excess supply. Our example shows that at a price of $5 per bag of maize, supply (216 000 bags) would exceed demand (150 000 bags). A surplus would remain which no one wanted to buy. Sellers would have to drop their price and supply less maize to clear the market. Again, the Government may want to fix the price above the equilibrium, in order to help farmers. To do this the Government would have to buy the surplus production. It may then have problems in finding the money and disposing of the surplus. A shift in supply will cause the equilibrium to change. For instance, the increase in crop yields after a good growing season, generally causes a fall in price, for example from E to F in Fig 2.5 Then in a drought year when yields are poor, supply falls and price rises again. These variations in prices between different years could be reduced by storage. Sometimes the Government may organize the storage. It can then aim to buy grain cheaply in good harvest years and to sell it in poor harvest years when the price is high. This way the costs of storage may be covered by the difference in grain price. By reducing the supply to the public in good harvest years and increasing it in poor years, the size of the shifts from year to year is reduced. This should in turn, reduce the size of the change in prices from year to year.

The size of the change in price, caused by a supply shift, depends on the elasticity of demand. This is shown in Fig 2.5. The first diagram (a) shows highly elastic demand. The slope of the demand curve is flat. When supply increases from S_1 S_1 to S_2 S_2 the price rises only a little. In the second diagram (b) demand in inelastic, so the slope of the curve is steep. Now the increase in supply causes a much bigger change in price.

The total revenue of all suppliers, in each case, is the same as the total spending by buyers. It is equal to the quantity sold multiplied by the price; the shaded rectangle in each case in Fig 2.5. When demand is elastic, total revenue is higher when supply is increased. When demand is inelastic total revenue is higher when supply is lower. Since demand for

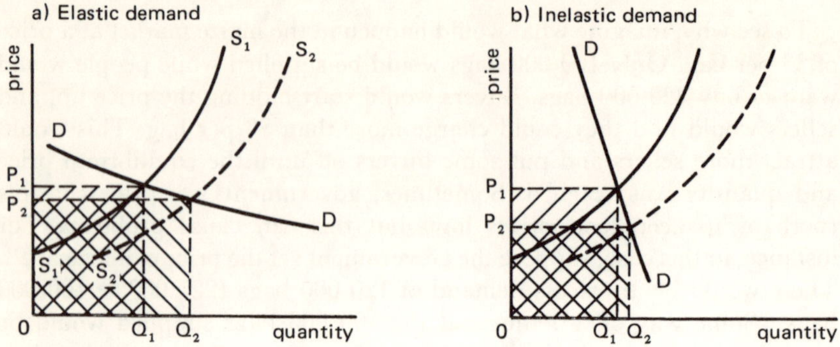

Fig 2.5 Demand elasticity and price changes

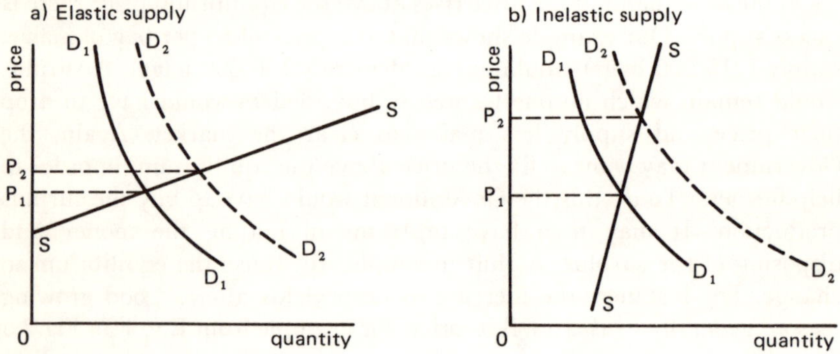

Fig 2.6 Supply elasticity and price changes

food is thought to be inelastic, farmers' income could be raised by restricting supply. Thus governments may control imports of food to help raise farmers incomes. Also farmers' organisations or marketing boards may try to limit the amount sold so as to raise prices. Of course these policies raise the cost of living for the buyers.

When there is a shift in demand, the effect on price depends on the elasticity of supply. This is shown on Fig 2.6. If, as in Fig 2.6b, supply in inelastic, prices change a great deal. Since, both supply of (and demand for) agricultural products is fairly inelastic fairly small shifts in demand (and supply) cause big changes in price. For this reason, agricultural prices vary quite a lot. They are said to be **unstable**. Government policies may be aimed at making farm prices more stable, as suggested above. In the long run, as population grows and incomes rise, so too does the demand for food. If supply remains inelastic, food prices will rise, perhaps quite fast. As food prices rise, so too does the demand for factors of production, such as land. Since the supply of farm land is inelastic, land rent may rise rapidly.

3 Marketing

3.1 The Importance of Marketing

The marketing task involves transferring goods from the producer (the farmer) to the consumer (housewife or other final buyer) in the right place, form and time. It also involves passing information about market demand back the other way from consumers to the producer. In village market places where farmers sell food crops to their neighbours, there is direct contact between producer and consumer. As towns expand and export trade develops, farmers and consumers become increasingly separated from each other and produce has to be transported, sometimes over long distances.

As trade grows, the marketing chain gets longer. Produce may be bought and sold by several traders or middlemen before it reaches the final consumer. There are three main types of traders. First there are the country buyers who buy from farmers either on the farm or in village markets. They may work for themselves or as agents for bigger organizations. Second there are wholesalers who deal only with other traders. They buy from country buyers and sell to the third type of trader, the retailer. He sells directly to the consumer from a shop or supermarket, a market stall or by hawking produce round the towns. Large private companies, co-operatives or marketing boards may be involved somewhere in the chain. Some possible alternative marketing chains are shown in Fig 3.1.

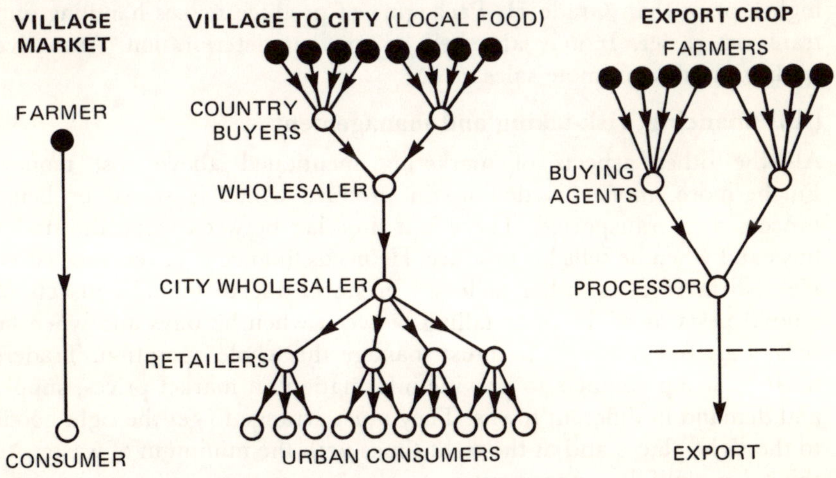

Fig 3.1 Alternative marketing chains

Marketing does not consist only of selling produce. It may include some or all of the following:

(i) Assembly and bulking of produce, transport, distribution and retailing

All of these are involved in moving food from producers to consumers. Value is added by supplying goods at the right place for sale. Because production is spread over many small scattered farms, produce must first be assembled and bulked. This makes transport and handling cheaper and more convenient. Often foods are transported over long distances to the main market centres. Here goods must be distributed to the retailers, who then sell to the consumer. Retailing involves presenting the goods attractively and perhaps advertising to attract customers.

(ii) Storage, processing, sorting or grading and packaging

Storage is almost certainly needed at some stage of the marketing chain. Traders can only provide a steady, regular supply to their customers if they have reserves in store. Value is added by supplying goods at the right time for sale. Farmers too may get a higher price by storing produce at harvest for sale later. Processing too may be done on the farm for some products, such as most food crops. Processing of other crop products, such as fibre, oil seeds, or a beverage such as coffee is better carried out by specialists on a large scale. Similarly meat animals are usually slaughtered and cut up by specialist butchers. Sorting and grading according to size, shape and quality makes trade easier. If grades are standardised, a buyer can order produce of a certain grade without even needing to see it first. Quality may affect the price received: for instance Grade I cocoa sells for a higher price than Grade II. Packaging of produce makes handling and transport easier. It may also help to prevent deterioration. Attractive packaging may promote sales.

(iii) Financing, risk-taking and management

All the other aspects of marketing mentioned above cost money. Furthermore, money is tied up in produce which is stored or being processed or transported. There is a time lag between when the trader buys and when he sells his produce. He needs finance to cover these costs. He also takes risks either of losses in store, due to moulds, insects or animal pests or of the price falling between when he buys and when he sells. Last but not least he must manage the whole operation. Traders must build up contacts to provide information on market prices, supply and demand in different towns. They must arrange to get the right goods to the right places and at the right times with the minimum of waste. All this takes skillful management.

The development of a marketing system is closely tied up with the development of the whole economy. Some countries have relied on agricultural exports for development. This usually means relying on thousands of small producers. A good road network and a well developed marketing system is needed for success. At the same time, industrial development leads to the growth of towns. The demand for food increases in the towns, which attracts traders. This causes the agricultural marketing systems to develop. Looked at the other way round, it seems that industrialization and specialisation can only take place if there is an effective marketing system. This allows the farmer to sell his produce for cash which he can then spend on a wide range of manufactured and other goods. His range of choice is widened.

3.2 Marketing Margins

The marketing margin for a particular item is the difference between the price the consumer pays and the price the producer receives. This sum of money covers the marketing cost including a profit for the traders. There is a widespread view that the price consumers pay for food is too high and the prices farmers receive are too low because marketing margins are excessive. It is even sometimes claimed that middlemen exploit the farmers, but it would be very difficult to prove. Of course, if the marketing margin could be reduced, consumers would pay less while the farmer would receive more, as shown in Fig 3.2. In this diagram the D_c curve

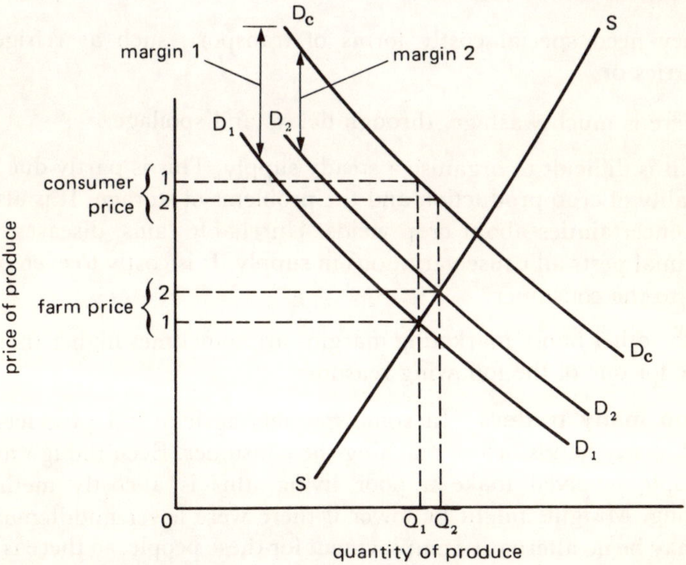

Fig 3.2 Marketing margin and prices

represents the consumer demand curve, faced by the traders. It the traders receive marketing margin 1, then the demand curve faced by the farmers for their product is given by the D_1 curve. Supply is equal to demand at quantity 1 and farmers receive price 1. If the marketing margin can be reduced to margin 2, the demand curve for farmers' products is the D_2 curve. Now supply is equal to demand at a higher quantity Q_2, and a higher farm price. At the same time the price consumers have to pay is reduced, so **both** producers and consumers would benefit from the increased sales and improved prices. However, it may be very difficult to reduce marketing margins, given the many costs involved. As people get richer they may even expect better processing and packaging of their food and be willing to pay for this. Marketing margins then rise in relation to the farm price.

There are several reasons why the cost of marketing agricultural produce is high compared with the cost of marketing an industrial product such as cement. First, agricultural produce comes from many small, scattered farms. Some of these are quite distant from towns or even from main roads. This means that the costs of collecting and transporting are high. Second, many agricultural products are bulky in relation to their value and highly perishable. Many fruits and vegetables as well as meat and dairy products cannot be stored for long without special treatment or processing. This means either;

(i) they must be processed on or near the farm, which may raise special problems and is costly **or**,

(ii) they need special costly forms of transport, such as refrigerated lorries **or**,

(iii) there is much wastage, through delays and spoilage.

Third, it is difficult to organise a steady supply. This is partly due to the seasonality of crop production and the problems of storage. It is also due to the uncertainties about crop yields. Unreliable rains, diseases, insect and animal pests all cause variations in supply. It is costly to even out the supply to the consumer.

On the other hand, marketing margins are sometimes higher than they need be for one of the following reasons:

(i) Too many traders In some markets agricultural products pass through many hands before reaching the consumer. Even though most of the people involved make a poor living, this is a costly method of marketing. Margins might be lower if there were fewer middlemen, but there may be no alternative employment for these people, so there is a real policy problem for governments.

(ii) Too few traders At some stages of the chain, a single trader may be the only buyer or seller of a certain product. He has a monopoly and can control the market to some extent. If the demand for the goods he sells is inelastic, he can make more money by restricting the quantity he supplies below the equilibrium level. This possibility is shown in Fig 2.3. Similarly, he can buy more cheaply if he restricts the quantity purchased below the equilibrium level. So by restricting purchases and sales, a monopoly trader can raise the margins he earns. However, in most African markets there are many traders, all competing with each other for purchases and sales. So it is difficult for any one person to form a monopoly and control the quantities traded.

(iii) Speculators Some traders buy goods when they are plentiful and the price is low, store them, then sell later when supplies are short and the price is high. Costs and risks are involved but the traders may make a profit. They are called speculators. In this case everyone benefits because consumers get a more even supply at a less variable price. Sometimes speculators can raise marketing costs, however. They may buy when supplies are short and the price is high, because they expect the price to rise even higher. Then, of course, they help to cause shortages and to drive the price up, by hoarding. If new produce comes on the market they may lose money, so there is a limit on the effect they can have.

Marketing is also involved in transferring inputs, such as high yielding seeds, fertilizers, sprays, animal feeds and medicines, from the producers, which may be large commercial companies, to the farmers, who in this case are the consumers. Now costs of distibution are high because farms are scattered and each one only needs small quantities of inputs. Yet local availability of supplies and equipment is essential for agricultural development. Many farmers would use fertilizers regularly if they could obtain them easily and at the right price. Very often their use is limited by poor distribution systems. Improved input marketing is needed to help farmers.

3.3 Co-operatives

A co-operative society is a business organization based on group activity. Members hope to benefit from sharing their resources and trading as a group. They contribute roughly equal shares so that no individual member can dominate the business activity. The aim is to provide services at cost rather than make a profit. If any trading surplus is earned it is added to the reserve funds of the co-operative. Each member has one vote in electing officers. However, most co-operatives are run by hired managers.

Agricultural co-operatives are set up for various purposes. Indeed many are multi-purpose co-operatives, which provide education and advice, credit and farm inputs as well as processing and marketing services for their members. Traditionally, farmers have often worked together under various mutual help schemes, without any formal organization. Formal agricultural co-operatives were first introduced in West Africa over 50 years ago for producers of cocoa and other export crops. These provided processing and marketing services and quality grades of exports improved rapidly. Co-operatives for food producers have been introduced more recently.

Marketing co-operatives are set up because local traders do not provide a good marketing service or charge too much. Of course, if the co-operative is to provide a better service at a lower cost it must be well managed. There may be price advantages in bulk selling of products and bulk buying of inputs. However, co-operatives also have to cover their operating costs and hired managers do not always control costs very effectively. Some lack experience, while others are more interested in their own careers rather than the benefit of their members.

Governments generally encourage agricultural co-operatives to promote improved marketing. Co-operatives are seen as democratic organizations aimed at common service rather than private gain. For such reasons many governments provide assistance to farmers in forming co-operatives. Co-operative colleges are set up for training managers and the co-operative movement may be given legal status. Organization is often based on a three tier structure. Primary societies operate at the village level. Groups of village societies form federated unions at district or regional level. Elected representatives from village societies attend union meetings. The apex society operates at national or state level. The structure is shown in Fig 3.3.

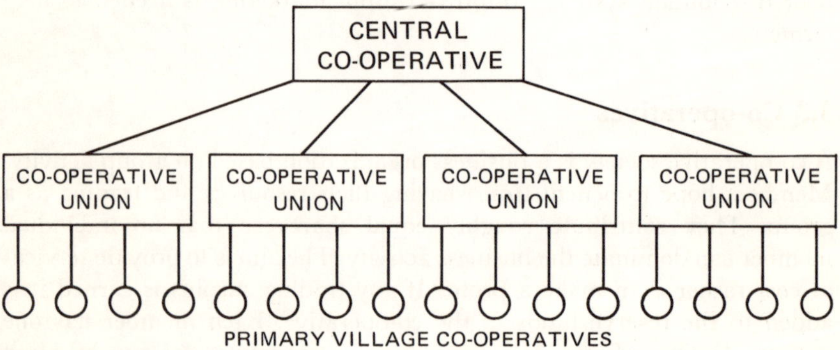

Fig 3.3 Co-operative structure

Co-operatives have had varying success in practice depending on two main factors, (i) the quality of leadership and (ii) willingness of members to co-operate. It is often claimed that behind every successful co-operative there is a selfless and dedicated leader. Linked with this is the need for good management. This is essential if the co-operative is to compete successfully with local traders, and gain the confidence and trust of the members. The easiest role for a new co-operative is to assemble a cash crop for direct sale to the marketing board. It has a sure outlet and no sales risk. To embark on further stages in marketing may require much capital and skilled management. Farmers are not always aware of the advantages of co-operation. Some may be tempted to sell privately if better prices are offered. Co-operatives do not always get the full support of members. Nonetheless the co-operative movement can provide an organization, through which farmers learn, over a period of time, to work together for their common benefit. It may be a good way for farmers to learn business methods needed for development of commercial agriculture.

3.4 Marketing Boards

A marketing board is a public body set up by the Government. The job of an agricultural marketing board is to control the market for a particular crop or livestock product. To do this, the board is given legal powers to regulate markets and fix prices. Its purpose is not to make profits, as a private trader might do. Rather the aim is to improve marketing services, such as grading or standardising the product, processing, packaging, storage, transport, promotion, advertising and selling exports overseas. An important aim of many marketing boards is to stabilize supplies and prices. They are generally financed by deductions from the sale price received.

The extent of the responsibility for control of the market varies from one marketing board to another. For convenience we may divide them into three main types, although in practice a single board may combine several sets of responsibilities.

(i) Non-trading, regulatory boards

A marketing board may simply be responsible for market research or technical advice to producers and also for advertising. Its main aim is to promote and encourage sales. It may also develop and apply quality standards. Produce inspectors grade the product before sale or export. The Export Control Offices in some French-speaking African countries provide such a service. The marketing board may also be responsible for controlling the numbers of traders by issuing licences to authorised buyers.

(ii) Price stablization

The main duty of some marketing boards is to try to stabilize the price received by producers. For foodstuffs or other products used within the country, the board may also try to stabilize consumer prices. Where there are only a few large producers it may be possible to control supply by issuing production quotas. Only licensed producers are allowed to supply the product and for each one the amount is limited to his quota allocation. Of course, it is still possible that in bad years producers may fail to meet their quotas. Even so, the supply and hence the price should be less variable than under a free market system. Another possibility is for the board to engage in trading and storage, to a limited extent. It may be able to handle and store, perhaps 20 per cent of the whole crop. A minimum guaranteed price is offered, and the board only buys when the market price falls below this level. It aims to sell at a time, when the market price is high enough to cover the costs of storage and management.

(iii) Monopoly trading boards

In many cases the marketing board is made a monopoly by law. All produce must be sold to the marketing board, which then becomes the only supplier for processors or the retail trade. There are some differences between export marketing boards and those for local markets with regulated retail prices. The marketing boards for Nigerian export crops (now replaced by commodity boards) were supposed to handle all the produce, even that for local consumption. An export board buys produce from farmers through its buying agents and sells on the world markets. It does not need much storage capacity unless it tries to influence the world price. The price the farmer receives can be stabilized using the board's reserve funds. When world price is high, the board pays the farmer a much lower price and builds up these reserve funds. When world price is low, the price the farmer gets may be subsidised. Reserve funds have been used for other purposes in the past so in effect the farmer has been taxed through the marketing boards.

Control of the market for a locally consumed product is much more complicated. Producers of local foods are often more widely scattered than cash crop producers, and the product once purchased must be redistributed to many small retailers and consumers. It is unlikely that the board can control trade in remote areas. To stabilize supplies and prices the board must either manage an overseas trading policy or establish buffer stocks. The first option means exporting surpluses and importing during periods of shortage. This requires careful management to avoid trading losses and over-dependency on world markets. The second option means storing sufficient food during periods of surplus supply, to meet shortage at other times. Buffer stocks require careful management to avoid major losses from trading and storage. Such policies are impossible for perishable products.

A general assessment of marketing boards is difficult. They have often proved useful in implementing government policy. Improved quality standards have been achieved and export prices raised. Services, such as processing of some products, can only be provided on a large scale. Marketing boards can help in this respect. The setting of guaranteed prices can reduce risks for the farmer and give him more confidence to produce for sale.

Against this, marketing boards have not always succeeded in their aims and some have proved very costly. A monopoly organization run by relatively few professionally trained managers is not necessarily more successful then traditional markets with many traders. Because there is no competition and the marketing board can charge a fixed levy on prices, too many staff may be employed and costs may drift upwards. Several studies have shown that marketing margins are higher under a marketing board, than under traditional trade. Price stabilization policies have sometimes broken down because of bad forecasting and the loss of reserves. Occasionally the guaranteed price has had to be dropped in mid-season. Skilled management is essential if marketing boards are to make a real contribution to agricultural marketing.

3.5 Government and Improved Marketing

An efficient marketing system is a key factor in the development of commercial agriculture and the rest of the economy. The Government has a duty to encourage and promote efficient marketing. There are various ways it can do this.

(i) General trade and pricing policies Many general policy decisions, such as whether to encourage foreign trade or how to control inflation, affect relative prices of agricultural goods. They affect agricultural markets. The farmer wants to know that the price of his products is not falling in relation to the general cost of living and the cost of his inputs is not rising.

(ii) Provision of communications Clearly produce cannot be traded between regions unless there are roads or rail links. The construction of feeder roads is one of the most effective ways of encouraging agricultural marketing. At the same time producers and traders need information on demand and prices in different market centres. The transmission of information is helped by improved postal and telecommunication services.

(iii) Provision of market facilities National or local Government can help by providing market places, in villages and towns. Market stalls and stores are needed, and perhaps assistance in providing for processing of crop or livestock products. In addition, a legal framework is needed to ensure that contracts are fulfilled, that standard weights and measures are used and products are not adulterated. Governments can make rules to prevent the establishment of monopolies and should consider providing a market information service. Producers and traders could benefit from advice on good marketing. Regular information on prices of main goods in all the main market centres, would help traders to know where to move goods and even out supplies.

(iv) Marketing institutions Governments may encourage the establishment of co-operatives or may set up marketing boards to control the market, as already discussed. Inputs may be distributed to farmers either through co-operatives or marketing boards. In any case it may be useful to establish rural services centres. These service centres can be built around a trading store, where farmers can obtain their input supplies and perhaps sell their produce. A machinery hire and service station might also be set up and a sub-branch of the agricultural credit bank.

3.6 Marketing Management on the Farm

Marketing is an important part of successful farm management. Increased production is only worthwhile if it can be sold for more than it costs to produce. Time and effort must be given to marketing off the farm. There are three main aspects.

(i) Collecting information Before starting production a farmer really needs to know where he can sell his produce at an acceptable price. He should also know the best time of year to sell, or how price varies over the year, and whether processing or packaging is required before sale. In order to choose between different alternative enterprises on the farm he needs this information for all possible items he might produce. But it is not really sufficient to know about the demand in the local markets. Better prices might be obtained by selling in more distant markets. Certainly big price differences occur between different towns in the same country and farmers often do not know where they can get the best prices. Similarly the farmer needs to know where and when he can buy the necessary inputs for his farm. Finally it is not enough to know the present situation. Forecasts of future demand and possible shifts of demand are also important for long-term planning. Time is well spent in collecting market information since it is valuable in guiding production and marketing decisions.

(ii) Marketing management Good marketing management, for the farmer, means getting his produce to market at the right time and place and in the right form to earn the best possible returns. This means organizing storage, processing, packaging and transport effectively and with as little waste as possible. Again efficient organization of the ordering and collecting of farm inputs is important for successful farming.

(iii) Entering marketing agreements Some farmers may wish to employ others to undertake some of the responsibility and risks of marketing. One possibility is to employ sales staff, commission agents or brokers to help sell produce. The farmer has to meet the cost but this may be recovered if a better price is obtained. Another possibility is to enter a contract with a particular buyer, before starting production. The contract might be to provide a specific quantity of produce at a particular price. The price might turn out to be lower than that in the open market but the farmer benefits by reduced risk and worry over his marketing operation. Often a big organization, such as an animal feed mill, is happy to arrange contracts with its input suppliers. Finally a farmer may encourage his neighbours to join him in forming a marketing co-operative as discussed above.

4 Farm Records and Accounts

4.1 The Value of Record Keeping

Record keeping is essential for successful farm management. Only by checking on input use and crop and livestock yields can a farmer correct faults and really control his farming operations. Simple facts may be noticed and remembered but written records are more reliable. They can also be passed from one person to another, for instance from the field worker to the farm manager.

Records together with accounts, which are simply financial records, are useful.

(i) In day-to-day management Checking on stocks of feed, fertilizer or chemicals, sudden emergencies such as disease outbreaks and whether bills have been paid,

(ii) In farm planning Since information on past performance can be used to estimate future results,

(iii) For financial management Particularly in applying for loans, (see Chapter 7),

(iv) For government planning purposes Which is why farm surveys are carried out.

For records to be really useful it is important to keep them up to date regularly. Some items, such as milk yields or egg production, should be recorded daily. Crop yields should be recorded at harvest time. Records must be accurate as possible. For instance it is better to actually weigh quantities of food or seed used and crop yields, rather than to estimate the amounts. Obviously records are only worthwhile if they are used. They should be summarised, studied and compared to guide decision making.

4.2 Inventories

An inventory is simply a list of live and dead stock on the farm at a particular date. Livestock includes not only the farm animals but also standing crops such as cocoa or oil palm and even annual crops such as maize if they have not been harvested. The dead stock consists of the durable items such as buildings and machines as well as the working capital stores of seed, food, fertilizers, fuels and chemicals. The inventory lists the quantity or number of each of these items.

The inventory must be up dated from time to time as items are bought, sold or used up. It then provides a useful check on stocks of farm inputs. The whole inventory should be carefully checked once a year on a particular date. The date should be chosen between the end of one harvest and the beginning of the next cropping season.

For accounting purposes, all the items in the inventory must be valued. This means the value or price must be estimated. It is fairly easy for stocks of food, seed or fertilizer. They can be valued at the purchase price in local markets. For items of durable capital it is more difficult. The value of those which are produced on the farm, such as drainage ditches or growing tree crops must be estimated, since there is no market price. Past records may be useful in estimating the cost of production. Even so, personal judgement will also be involved. There is also the problem of estimating depreciation. As we saw in Chapter 1, this problem applies even to purchased items such as machines.

The value of a second-hand machine depends on both its age and its condition. Its condition will be affected by the amount of work it has done and how well it has been maintained. However, to keep the calculation simple, depreciation is usually based on age only. Using what is called the **straight line method**, annual depreciation is simply the average loss in value per year. For example consider a sprayer which cost $50 when new, which is expected to last five years before it needs replacing. The annual depreciation is:

$50/5 = $10

If, at the end of the five years, it would have a second-hand value of $50 the depreciation would be less:

$(50−5) /5 = $9

The value of any durable item can now be estimated by subtracting the annual depreciation from the previous year's value.

Some items, particularly growing trees and livestock, actually gain in value over time. Again judgement may be needed to estimate these gains. They are important and should be recorded, because they represent part of the farm output. The total value of all the items on the inventory is known as the **valuation**. One taken at the start of the production period is called the **opening valuation** and one at the end of the period is the **closing valuation**. Of course, the closing valuation for this year can also serve as the opening valuation for next year.

4.3. Farm Accounts

It is advisable to keep a record of all financial transactions. Records of both sales and purchases can be kept in the same account book. Usually,

Table 4.1 Simple Account Book

SALES AND RECEIPTS					*PURCHASES AND EXPENSES*			
Date	Details	$	c		Date	Details	$	c
14 Jan	7 bags of maize (to John Oni-trader)	37	50		4 Jan	Labour hire (Thomas Ade 3 days)	14	60
26 Jan	3 dozen eggs (village market)	5	50		18 Jan	Seed Maize (1 bag from Government seed unit	8	20
					20 Jan	Cutlasses (10 from village store)	28	90

(and so on for the rest of the year)

sales and receipts are recorded on the left hand page, while purchases and expenses are noted on the right hand page as shown in Table 4.1. The important facts to note, about each transaction, are the date, what the item was and the sum of money involved. It may also be useful to record the name of the buyer or seller.

It is convenient to start using a new account book each year, on the day of the opening valuation. At the end of the year, the total receipts from sales of output and the total expenditure on purchases of inputs, can be added up. The difference between total receipts and total expenditure can be calculated, to give a measure of the **cash flow**. In some years, especially when big capital investments are made, the cash flow may be negative. This means that total expenditure is bigger than total receipts. On a successful farm, receipts are bigger than expenditure in most years.

The cash flow is **not** a measure of profit. To estimate profit two sorts of corrections are needed:

(i) Differences between the opening and closing valuations must be taken into account

If the value of the farm has increased, this amount must be added to the cash flow. Where a big capital investment has been made the increase in valuation will cancel out the negative cash flow. Alternatively if the value of the farm has decreased, the cash flow should be reduced accordingly. In either case, the closing valuation should be added to receipts and the opening valuation should be added to the costs.

(ii) Home consumption must be included

Any produce which is used in the home, is really part of the farm output. It should therefore be recorded and valued at the going market price. The total value of home consumption can then be added to the cash flow.

It is sometimes suggested that the value of unpaid family labour should be subtracted, in calculating farm profit. Similarly, the rental value of land could be subtracted, even when no rent is paid. We do **not** recommend these ideas. The values can only be guessed, so it is probably better to ignore these costs. The summary of the trading account may then look like Fig 4.1.

Many farmers will wish to analyse their accounts in more detail. For instance they may wish to keep separate records of sales for different enterprises. The best method then, is to use a **cash analysis account book**. A suitable account book can be bought from a stationers or a farmer could make his own (see Table 4.2.). Apart from columns for the date, item and total cost, additional columns are ruled for each main item of income or expenditure. At the end of the accounting year it is then quite easy to add up the total for each item. As a check on the arithmetic, the

Table 4.2 Cash Analysis Account Book

SALES AND RECEIPTS						*PURCHASES AND EXPENSES*						
Date	Details	Total	Maize	Poultry and eggs	Tools and equipment	Date	Details	Total	Seeds	Fertilizer	Labour	Tools and equipment
		$ c	$ c	$ c	$ c			$ c	$ c	$ c	$ c	$ c
14 Jan	7 bags of maize (to John Oni-trader)	37 50	37 50			4 Jan	Labour hire (Thomas Ade 3 days)	14 60			14 60	
26 Jan	3 dozen eggs (village market)	5 50		5 50		18 Jan	Seed maize (1 bag from Govt. seed unit)	8 20	8 20			
						20 Jan	Cutlasses (10 from village store)	28 90				28 90

(and so on for the rest of the year)

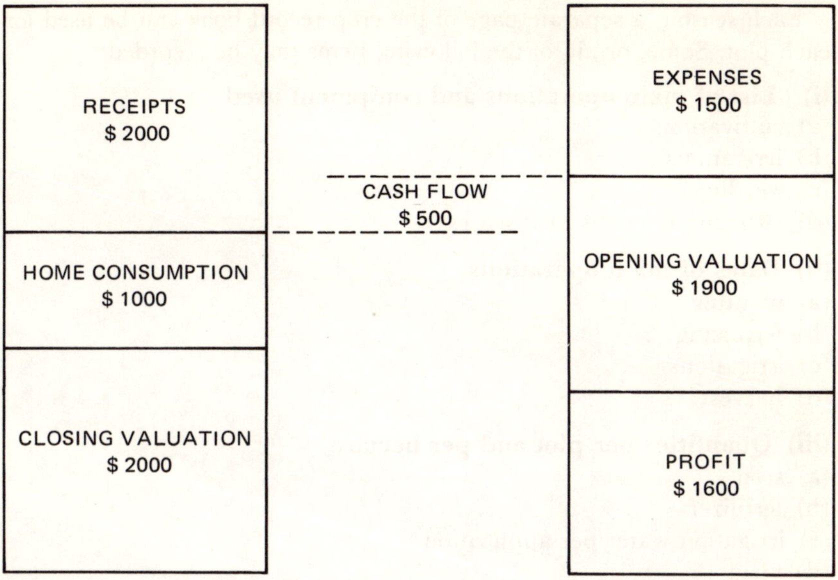

If a loss occurred it would appear on the left-hand side

Fig 4.1 The Trading Account

sum of the item totals should add up to the grand total for sales or purchases.

Now the inventory valuations must be split up item by item. This allows the output to be calculated for individual enterprises. For example, the output of a poultry enterprise is then the total sales of eggs and birds **plus** any increase in the value of birds between the opening and the closing valuation. Similarly, the output of maize is the total amount sold plus any increase in the quantity stored. In the case of inputs, such as seed, we have;

Cost = Purchases + Stocks at opening valuation − Stocks at closing valuation

4.4 Crop Records

Production records are used to keep a check on the physical quantities of inputs used and the yields obtained. Crop records are more useful when they are related to the area grown. For example, yield per hectare is a more useful measure than total yield. First of all, then, a farm plan should be drawn with all the plots marked. The area of each plot should be measured and recorded.

Each season, a separate page of the crop record book can be used for each plot. Some, or all, of the following items may be recorded:

(i) List of main operations and equipment used
(a) cultivations
(b) irrigations
(c) weeding
(d) spraying for pests or diseases

(ii) Dates of main operations
(a) planting
(b) fertilizing
(c) irrigations
(d) harvest

(iii) Quantities per plot and per hectare
(a) seed
(b) fertilizer
(c) irrigation water per application
(d) yield

(iv) Crop disposal – quantities
(a) stored
(b) consumed and
(c) sold (with dates)

It may also be useful to record the costs of inputs and the price received for crop sales in the production record book. But this information should be available from the cash analysis account book.

Production records for permanent crops are similar to those for annual crops. But some trees are widely scattered, instead of being grown in a plantation. Then it is better to record inputs and yields per tree, instead of per hectare.

4.5 Livestock Records

It may be better to keep livestock records on cards, rather than in a book. Many animals and poultry are housed in sheds or pens where the record cards can be pinned to the wall. Generally, a separate record card would be kept for each pen of fattening animals or poultry. But for breeding sows or milking cows it may be better to keep a separate record for each animal.

The main input records for livestock are; (i) types and quantities of food fed and (ii) health records; vaccinations, dippings and so on, with dates. It is also important to keep a regular check on numbers. This helps in control of losses due to sickness or thefts. Also it is needed for calculation

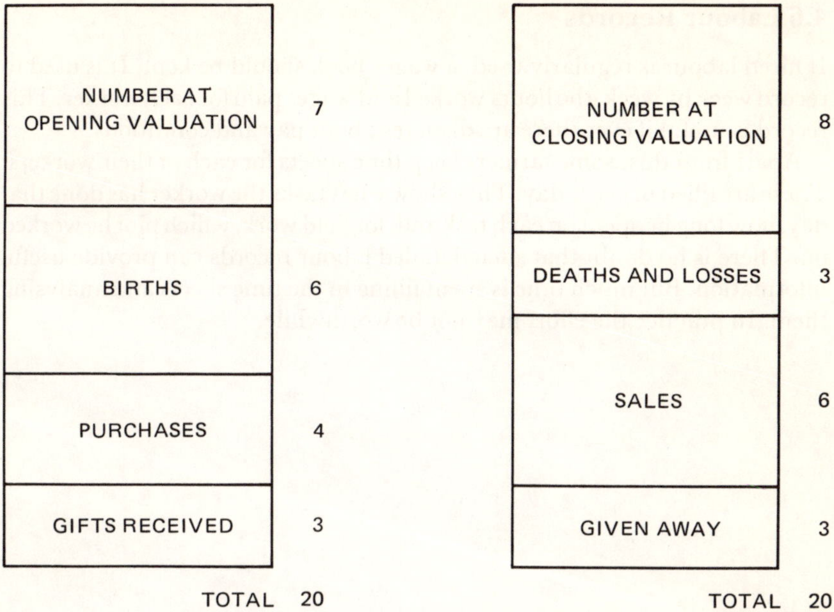

Fig 4.2 The balance of livestock numbers

of yield per animal. Over a certain period, there will be increases in numbers due to births, purchases and perhaps gifts received. Over the same period numbers are reduced by accidental deaths, thefts, slaughterings sales and gifts. The number at opening valuation plus additions to numbers over the year should equal the number at closing valuation plus those lost, or given away. (See Fig 4.2.)

Production records vary according to the type of livestock.

(i) Fattening stock At least the time from birth to slaughter should be recorded, for individual animals or for a batch. A better check on growth rates can be kept if the animals are weighed regularly. They can then be sold or slaughtered at just the right weight to get the best price. The time and cost of regular weighing may be worth while

(ii) Breeding stock A record should be kept for each female, of the date of each pregnancy, the date of birth of the young and the number born. Such records are useful in choosing which females to keep and which to cull. Records also help in maintaining a regular breeding cycle.

(iii) For dairy cows or laying hens it is usual to keep daily records of the yields. The daily milk yield may be recorded for individual cows, to guide feeding and breeding policy. For poultry the total yield of good and cracked eggs should be recorded for the whole shed.

4.6 Labour Records

If hired labour is regularly used, a wages book should be kept. It is used to record week by week, the hours worked and wages paid for each worker. This record is useful if ever there are disputes about pay and conditions.

Apart from this, some farmers keep time sheets for each of their workers. These are filled in every day. They show what tasks the worker has done that day, how long he spent on each task and, for field work, which plot he worked on. There is no doubt that such detailed labour records can provide useful information. But much time is spent filling in the time sheets and analysing them. In practice the effort may not be worthwhile.

5 Farm Business Analysis and Planning

5.1 Enterprise Gross Margins

The contribution that each enterprise makes to the total farm income is known as the **enterprise gross output.** It is the total yield, or physical output, multiplied by the expected sale price. This method of estimation is used, even if the produce is to be eaten at home. Crop returns are usually worked out on a per hectare basis, though for oil palms and other permanent crops the return per tree may be used. Several products may be obtained from one enterprise, for instance, palm fruit and palm wine or grain and straw, or several crops in a mixture. The entreprise gross output includes the value of all the products. It may also include the gain in value of a permanent crop which is still growing.

The gross output per head of fattening livestock is usually measured as the gain in value. It is the sale price when mature minus the purchase price when young. If the growing period is longer, or shorter, than a year, the figures must be adjusted to give gross output per year. For breeding animals the gross output is the average number of young born per year times the value of each. Similarly the quantity of milk per cow, or the number of eggs per hen per year, times the price, gives the gross output.

Costs of production are also worked out on a yearly basis. They are of two kinds.

(i) Fixed or overhead costs These include full-time hired labour or field machinery. These costs apply to the farm as a whole, rather than to individual enterprises. They do not vary, even though the farmer may change his pattern of production. For instance, the costs of livestock buildings have to be met even when no animals are kept. The costs of land clearing are the same whether groundnuts or cotton are grown.

Fixed or overhead costs include rents if any, wages of regular and family labour, machinery depreciation (see Chapter 5) and repairs of buildings, terraces, ditches and so on, office costs if any.

(ii) Variable costs This includes seed or animal feedstuffs. These costs apply to a particular crop or enterprise. The amount of cost varies directly with the size of the enterprise. For instance, when a farmer decides to grow more groundnuts he must use more seed. If he keeps poultry he must buy poultry feed.

Variable costs include:
(a) For crops: seeds, fertilizers, sprays and casual labour hired for particular tasks, such as harvesting.
(b) For livestock: feedstuffs, vet's bills and drugs.

For any crop or livestock enterprise the gross output, minus the variable costs, gives the **enterprise gross margin**. It measures the contribution of that enterprise to the farm profit. The total gross margin from all the enterprises on the farm, minus the fixed costs, equals the farm profit. Since fixed costs do not vary with changes in the farming system, the larger the total gross margin, the larger is the farm profit (see Fig 5.1). Gross margin calculations are shown in Table 5.1.

Table 5.1 The calculation of gross margins

Cost and output for a 5 hectare farm with 3 hectares of upland rice and 2 hectares of maize

1. Fixed costs		$
Contract work for land clearing at $650 per hectare		3250
Regular labour force		1500
Tools and equipment – depreciation		500
Total fixed costs		5250
2. Rice Production (3 hectares)		$
Revenue 3 × 50 tonnes at $450 per tonne milled		6750
Variable costs	Seed	54
	Fertilizers	150
	Casual labour	100
	Milling costs	120
Gross Margin = revenue − variable costs =		6330
Gross margin per hectare = 6330/3 = $2110		
3. Maize production (2 hectares)		$
Revenue 2 × 7 tonnes at $120 per tonne		1680
Variable costs	Seed	42
	Fertilizers	140
Gross Margin = revenue − variable costs =		1498
Gross margin per hectare = 1498/2 = $749		
4. Farm profit = total gross margin − fixed costs = (6330 + 1498) − 5250		$2578

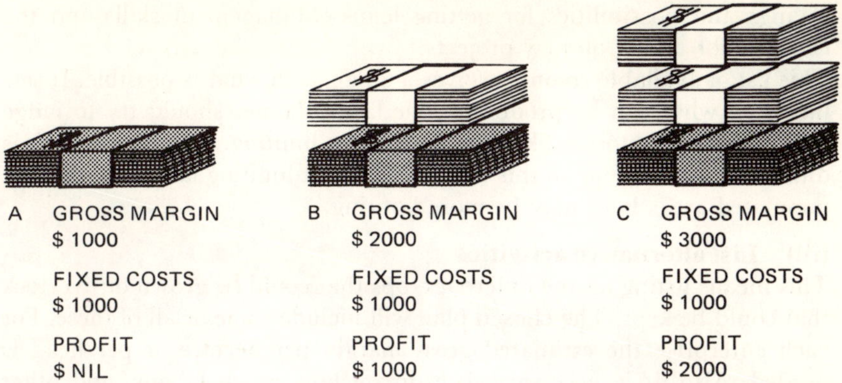

Fig 5.1 Gross margins and farm profit

5.2 Farm Planning

A successful farmer plans ahead. He plans his production system by deciding what to grow next year or even in several years'time. If he is a businessman he will want to maximise his profit. He will choose the enterprise (or combination of enterprises) which is likely to earn the largest gross margin. To do this he needs estimates of the average gross margins he can expect to earn, per hectare or per tree for crops and per head for livestock.

These estimates can be based on farms records and accounts from the past, if they have been kept (see Chapter 5). If a new enterprise is proposed, or if the farmer does not keep records, he may have to rely on:
(i) records from government research stations,
(ii) records from neighbouring farms,
(iii) estimates by an agricultural adviser.
It is important to assess the market prospects for new or extra products, The estimated gross margin should be based on average prices, expected in a *normal* season. If there are serious doubts about future yields and prices, it may be worth making several plans using different sets of prices each time.
Farm planning involves the following steps:

(i) Decide upon objectives
It is usually assumed that the main objective is to earn profits.

(ii) List available resources
Present land area and its use; additional areas available. Present labour force and its make up; hours available for work on the farm; availability of hired labour. Buildings, machinery and equipment on the farm, cash

savings and possibilities for getting loans. Management skills and the capacity for taking on new projects.

This list of available resources gives a guide as to what is possible. It sets limits on what can be produced. The farm planner should try to judge which of these factors is likely to be the most limiting. In some places it is difficult to hire labour so this may be the most limiting factor. In densely populated areas land may be more limiting.

(iii) List alternative activities

This means listing all the different crops that could be grown or livestock that could be kept. The chosen plan will include some or all of these. For each enterprise the estimated gross margin per hectare or per head is needed. Also, it is necessary to estimate how much labour, and other limited resources, may be needed per unit.

(iv) Prepare budgets

This means estimating the costs and returns of alternative plans. With estimates of the gross margins this should be fairly easy. A good general principle is to **choose the enterprise which yields the highest gross margin per unit of the most limiting factor**. If the land is most limiting then choose the crop which gives the highest gross margin per hectare. Pigs or poultry may be added without using more land. If labour is more limiting choose the enterprise which yields the highest returns per man-day.

(v) Choose a plan

The planner should have calculated the total gross margin that can be earned from different possible farm systems. If the estimated fixed costs are taken away, farm profit is left. When two or three plans are compared, the one which gives most profit may be chosen.

(vi) Put the plan into action

Success will depend on good management and supervision.

5.3 Planning Land Use

For several reasons it may be unwise to grow the same crop over the whole farm. First, there may be areas of different soil types suited to different crops. River valley bottoms may be suited to rice or vegetables. Uplands may be better suited to cereals. Some land may already have trees growing so it is not available for arable crops. Second, some land may be near the buildings or have direct access by road. This can be cropped heavily. More distant land may be cropped less frequently. Finally, a crop rotation may be needed to produce high yields. If cereals should only be

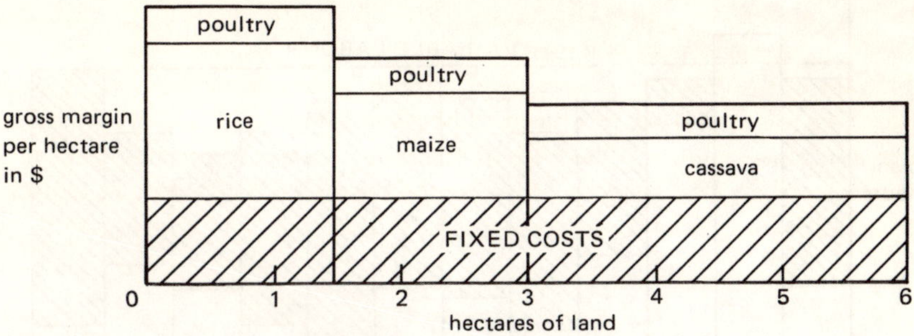

Rice limited to $1\frac{1}{2}$ hectares by lack of more suitable land
Maize limited to $\frac{1}{3}$ of rotational area
Poultry needs no extra land but adds to gross margin per hectare

Based on Upton, M. & Anthonio, Q.B.O. (1964) *Farming as a business* O.U.P. Figure 6.

Fig 5.2 Marketing return per hectare

grown five years in ten, then only half the land should be under cereals at any one time.

A drawing or map of the farm is very useful. Roads and buildings can be marked on it and areas of different soil types. If it is drawn accurately plots can be measured. The areas of crop rotation can also be worked out on the farm map.

When land is limited, the first choice of crop will be the one with the highest gross margin per hectare. If only part of the land can be used for this crop, the next choice will have the next highest gross margin and so on (see Fig 5.2). Cattle, sheep and goats also need grazing land. With estimates of the gross margin per head and the stocking rate per hectare, it is possible to calculate the gross margin per hectare. Sometimes these enterprises may give higher gross margins than arable crops. Pig and poultry enterprises can be started to raise the total gross margin without using any land directly (see Fig 5.2).

5.4 Planning Labour Use

Labour needs on the farm follow a seasonal pattern of peaks and troughs (see Fig 1.2). When labour is limited, the amount available at peak periods really limits what can be grown. For instance, the number of workers available at planting time may be the most limiting factor **not** the total labour availability. Then the best enterprise to choose is the one which gives the highest gross margin per hour of labour at planting time.

Having chosen a main enterprise, it is often profitable to find another which uses labour at different times of the year from the first (see Fig 5.3). Total gross margin can be increased without raising the total labour cost. For labour use planning it is helpful to estimate a labour profile like Fig 1.2 for the proposed plan.

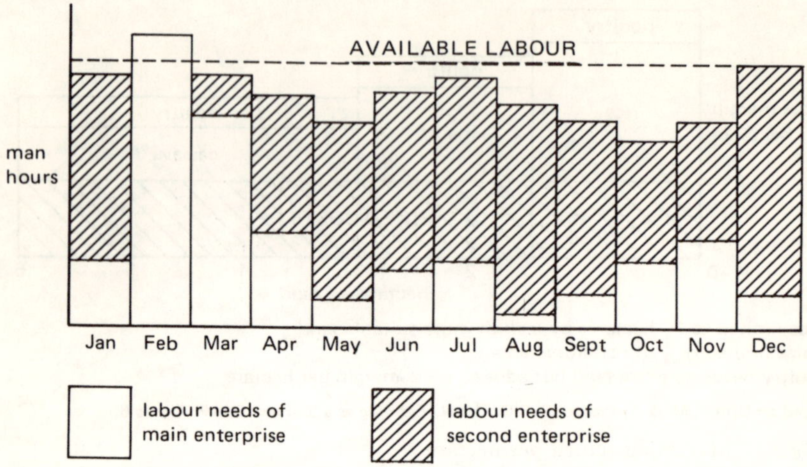

Fig 5.3 Better use of labour by combining enterprises

5.5 Capital Budgets

Very often the proposed plan involves capital investment. New buildings and equipment may be needed or land clearing and drainage. Tree crops may be established or breeding livestock numbers may be increased. In these cases gross margins do not show all the changes in costs and income. Cash flow analysis is needed.

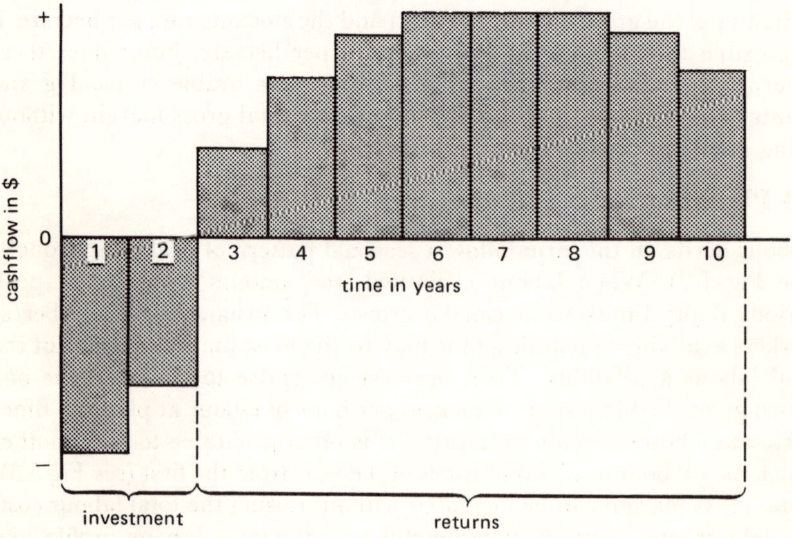

Fig 5.4 Cash flows over time

With any capital investment there is always a series of varying cash flows. Extra income received from sales is a **positive** cash flow (it adds to the farmer's income). Costs are negative cash flows (they reduce the farmer's income). In cash flow analysis the differences between total income and total costs are estimated for each period. For a very short-term investment, such as new seeds, the cash flows may be estimated for monthly periods over one season. For long-term investments, yearly cash flows may be used over, say, five years. Most investments start off with negative cash flows because of the costs of establishing the project. Positive cash flows come later when returns are produced (see Fig 5.4).

The first step, before estimating the cash flows, is to work out the timing of the project. – When will it be established? Will the establishment cost be spread out over several periods? How often will the equipment need to be replaced? Once the plan has been worked out in detail the total income from sales of produce and the total costs or expenses are estimated for each period. If any extra taxes will be paid these should be included as expenses, while any grants or subsidies can be treated as additional income. The difference between income and expenses in each period is the cash flow. Of course it is not necessary to estimate the cash flows for the whole farm. Only the **extra** income resulting from the planned project and the **extra** expenses are used to estimate the project cash flows. An example of cash flow estimates for a poultry project are given in Table 5.2.

Table 5.2 Cash flow estimates (for 5 year period) 250 bird, deep-litter poultry unit

Cash items	Initial investment $	Year 1 $	Year 2 $	Year 3 $	Year 4 $	Year 5 $
1. Income						
Eggs		2250	2250	2250	2250	2250
Sale of hens		250	250	250	250	250
Sale of second-hand equipment						200
Total income		2500	2500	2500	2500	2700
2. Expenses						
Buildings and equipment	1200					
Point-of-lay pullets	500	500	500	500	500	
Feed and drugs		750	750	750	750	750
Labour		150	150	150	150	150
Repairs		10	10	10	10	–
Total expenses	1700	1410	1410	1410	1410	900
3. Cash flow						
Income – expenses	–1700	1090	1090	1090	1090	1800

The proposed plan is to establish a 250 bird deep-litter poultry unit to produce eggs and culled hens. A survey of local markets has shown that there is a demand for both eggs and poultry meat which is at present being supplied from outside the area. It seems that the produce can be sold locally at the current prices. Although the project could be continued for many years into the future, the budget only covers the first five. It was decided that such a project should earn a reasonable return within five years, otherwise it would be too risky.

The initial investment at the start of the project consists of the costs of the buildings, equipment and the first batch of point-of-lay pullets. This is a negative cash flow. Each year all the birds are replaced so income is received from sales of birds as well as eggs, while the costs of a new batch of point-of-lay pullets is incurred. Annual running costs also include food, drugs, labour and an allowance for repairs. It is assumed that, in the fifth year, the project will be wound up so there is some extra income from sales of second-hand equipment and no cost of replacement pullets.

Once such a series of cash-flows have been estimated, it is then possible to calculate whether the capital investment is worthwhile. There are several different ways of measuring whether a project is worthwhile; the following are the most common.

(i) The payback period

The payback period is one of the simplest methods of judging a project. It is simply the length of time it will take to recover the initial investment. In the present example the pay-back period is between one and two years. The positive cash-flow of $1090 in year one is not sufficient to cover the initial investment of $1700 but by the end of year two the total positive cash flows of 2 x $1090 = $2180 will be more than enough. This is only a rough-and-ready measure of whether the project is worthwhile and there is no standard rule for deciding the longest acceptable payback period. In this case two years may be thought short enough to justify proceeding with the plan.

(ii) Return on average investment

It is more useful to estimate the rate of return on the capital invested, since this can be compared with the interest rate payable on borrowed money. One way of doing this is to estimate the annual depreciation of capital and subtract this from the annual cash flow to arrive at the annual return. The average of the initial investment and the capital value at the end of the project gives the average investment. The rate of return is then the annual return as a percentage of the average investment.

From Table 4.2 we find the initial investment is $1700, while the capital value at the end is $250 for old hens plus $200 for equipment, giving a total of $450. The calculations are then as follows;

1. Annual depreciation over 5 years $= \frac{\$1700 - 450}{5} = \$250$

2. Annual return = \$1090 − 250 = \$840

3. Average investment $= \frac{\$1700 + 450}{2} = \$1075$

4. Rate of return on average investment $= \frac{840}{1075} \times 100 = 78\%$

(iii) Discounting

The method just described only works where the cash flows are the same in each period after the initial investment. It does not take full account of the costs of borrowing. Discounted cash flow methods are generally more reliable but are more complicated. A full description of these methods is given in Chapter 8 of 'Farm Management in Africa' by Martin Upton published by Oxford University Press (1973). A credit advisor should be able to help with such calculations.

6 Agricultural Finance

6.1. Investment Saving and Growth

Economic growth and development depends upon investment in new capital. Increased investment leads to increased production and incomes. The amount of new investment depends, in turn, on the rate of saving. This represents a cost to society, because it means consuming less at the present time. There is a choice between a high standard of living now or a higher one in the future.

The circular flow of income, within a country is shown in Fig 6.1. This is, of course, a very simplified view. The left hand box, (labelled *firms*) represents all factories, farms and other businesses where goods are produced. These firms pay out wages, salaries and profits which add up to the total national income. The population live in households which receive this income. Part is spent on consumption of food and other items for which firms receive money. The other part which is not consumed is saved, and can be used for investment in firms.

Some farmers and other businessmen save in order to invest. Many people save for other reasons, in readiness for old age, for an expensive ceremony or just as a reserve in case of emergency. Savers have no way of knowing who wants to borrow for investment. Banks and other credit agencies provide the link between savers and borrowers. Savings then do not lie idle and businessmen can get credit for investment.

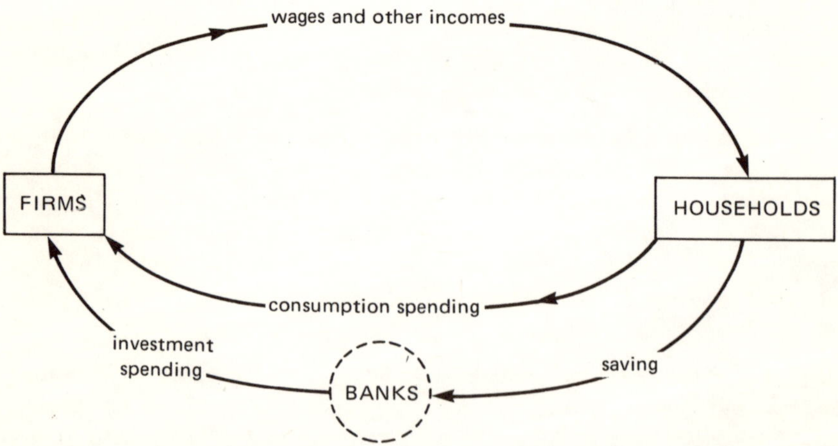

Adapted from Harvey, J. and Johnson, M. (1971) *Introduction to Macroeconomics* Macmillan

Fig 6.1 Money flows in the economy

Some important income flows have been missed out from Fig 6.1. First there is overseas trade. Money (foreign exchange) is earned from exports, while it is paid out for imports. Problems can arise when the cost of imports exceeds the value of exports. Second there is the Government. Income is withdrawn from firms and households in the form of taxes. The Government then spends this income in three main ways:

(i) Social services such as the police, the army, roads and schools.
(ii) Direct government investment in State Farms or Nationalised industries.
(iii) Encouragement and assistance of private producers through research and development, inputs, grants and loans.

In these ways the Government can influence the rate of economic growth and development. By raising taxes it forces people to save. The money can then be spent or invested to meet development aims. Governments may also raise investment funds by overseas borrowing. The scope is limited because it increases the national debt and causes dependency on other countries.

6.2 Investment in Agriculture

Agricultural development and increased farm production is essential for national economic growth.

(i) To maintain the national food supply As incomes rise and the population grows, the demand for food increases. If food production does not increase as fast, there are serious problems. Food prices rise so workers in the towns become discontented. Shortages may occur and some people may go hungry. Food imports grow and this is a drain on scarce foreign exchange. For all these reasons food production should be increased to keep pace with rising demand.

(ii) To provide employment in rural areas As wages and salaries in the towns rise, while farm incomes stay low, young people leave agriculture. They cannot all find jobs in the towns and many are left unemployed. Poverty and crime flourish and this can become a serious problem. When agriculture is developing and growing, young people can see the opportunities for making a good living in farming. They then stay on the farm and contribute to the improvement of village life.

(iii) To raise national income For many countries, agricultural exports are important in earning foreign exchange. Alternatively, foreign exchange is saved when farmers start to produce food and other goods which used to be imported. Agriculture also provides essential raw-materials for industry.

(iv) To provide a market for industrial goods As farm incomes rise, farmers spend more on their homes, clothes, furniture and all sorts of manufactured goods. Industrial manufacturers are encouraged to increase production if they know there is a growing demand in rural areas.

For these reasons governments have become more and more concerned to raise agricultural production. Some direct government investments have been made in ranches, tree crop estates and government farms. There is much more scope for raising agricultural production by encouraging private farmers. Governments can assist and support agricultural development in several ways:

(i) promoting efficient agricultural marketing (see Chapter 3),
(ii) financing agricultural research, plant and animal breeding and development of improved methods,
(iii) providing an agricultural extension service to spread the new ideas to farmers,
(iv) ensuring the local availability of supplies and equipment,
(v) increasing agricultural credit.

This last is unlikely to be successful unless the other four kinds of support are available.

6.3 Credit Agencies

Banks and other credit agencies, such as co-operative credit unions, village societies or private moneylenders, must manage their funds carefully. The total funds available for lending are limited and it is wise to keep a certain amount in reserve. A bank's funds are increased by savers, who deposit their savings in the bank. These people may want to withdraw their savings at any time, so the bank must have reserves to

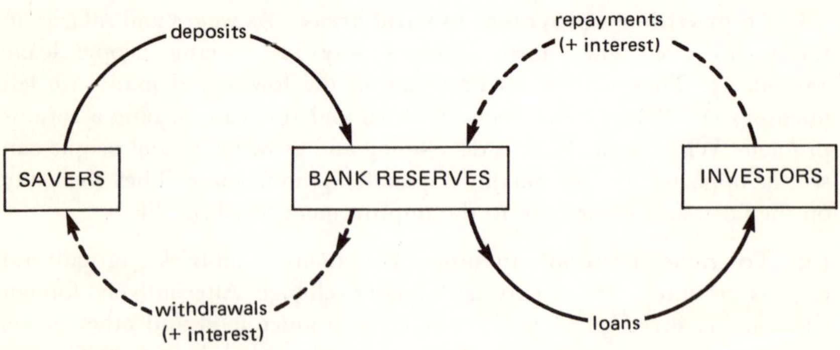

Fig 6.2 Money flow through a bank

meet this possibility. Money which is lent out should eventually be repaid, and this goes back into the bank's reserves. Overall, the bank needs to make sure that the outflows of money (deposit withdrawals and loans) do not exceed the inflows (new deposits and loan repayments). These flows of funds are shown in Fig 6.2.

The bank manager or any other credit agent, before making a loan, needs to know that the borrower is likely to repay. He needs some security. It can be provided in two main ways:

(i) Personal knowledge of the borrower and careful checking that he is creditworthy. It may be necessary to study his business accounts and budgets, and even to supervise his use of the credit. All this is costly for the credit agency.

(ii) Collateral security, meaning some possession of the borrower which the lender can keep if the loan is not repaid. Under a mortgage, land or buildings provide the collateral security. The same is true when land and tree crops are pledged. Machinery or equipment bought under hire-purchase is itself collateral security in case the payments are not made.

Interest is the price paid for the use of money. It is usually stated as a percentage of the sum owing and is paid annually. A man who borrows $100, at 10 per cent annual interest, owes $110 at the end of the year ($100 plus 10 per cent of $100). If he pays nothing back before the end of the second year, he will then owe $121. This is made up of the $110 owed at the start of the second year plus interest at 10 per cent. The actual interest cost rises (from $10 in the first year to $11 in the second) because the total debt rises (from $100 originally to $110 in the second year). Interest charged in this way is called 'compound interest'.

Savings which are deposited with a bank usually earn some interest but at a fairly low rate. In order to cover its costs, the bank must charge a higher rate of interest on its lending. The interest charged on loans must cover:

(i) the interest paid on deposits (the opportunity cost of funds),
(ii) administration and staff costs,
(iii) losses due to unrepaid loans (or a risk premium to allow for default).

These costs may vary from one situation to another. Even so, banks usually charge a fixed, standard rate of interest on all loans of a particular kind.

6.4 Farmers' Credit Needs

We have already seen that capital is needed for running a farm. Seasonal, working capital is necessary as well as more durable, buildings equipment

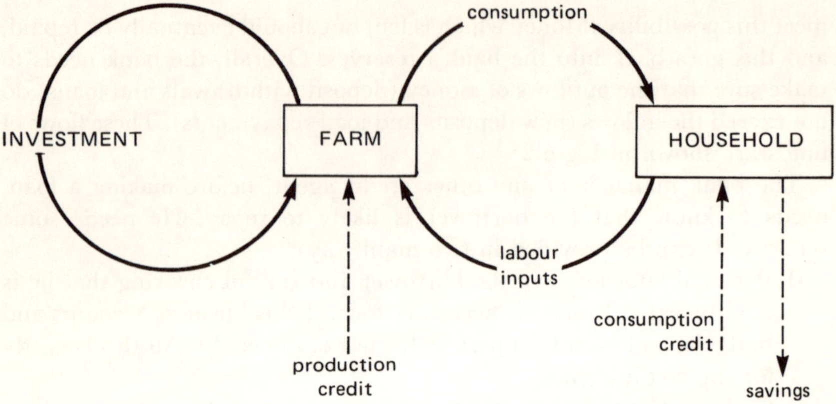

Fig 6.3 Farm income flow

and livestock. Extra investment in new crops, livestock and methods is vital for increasing farm output. To produce more, farmers must spend more. Capital is also needed for improved storage, processing and marketing.

Many farmers succeed without using credit. They provide all their capital investment needs from their own savings. Ways in which farmers can build up their own capital were discussed in Chapter 1. Fig 6.3 shows the circular flow of income on the farm. These are not all money flows. Food which is consumed in the household is not exchanged for money. But farm incomes vary from one season to another and so do household needs. At some times the family may be able to save but at other times credit may be needed.

Sometimes credit is needed for consumption; when the harvest fails or one of the family is sick and unable to work. Loans are sometimes used to pay for ceremonies. There are dangers in borrowing for consumption because the loan does not increase future production. It may be difficult to repay if there is no extra income produced.

Production credit, used for investment, should increase farm income. If the investment is worthwhile the return should cover the costs of repayment and interest on the loan and still leave some extra profit. The length of time needed to recover the cost varies. There are three types of loan:

(i) Short-term loans For purchase of working capital such as seeds, fertilizers, or labour hire. These are repayable within one or two seasons.

(ii) Medium-term loans Used to finance early maturing projects such as poultry and piggeries, crops which are not annual but which mature quickly, machinery and equipment. These are repayable in two to four years.

(iii) Long-term loans For the establishment of tree crops such as cocoa, oil palm or rubber and for breeding livestock. The repayment period is longer than four years.

Sources of farm credit may be classed as formal or informal. Informal sources of credit include: relatives and friends, traders and shopkeepers, produce buyers and moneylenders. The interest rates charged by informal sources vary widely. Relatives and friends may give interest-free loans while moneylenders sometimes charge up to 100 per cent.

The **advantages** of informal credit are:

(i) convenience to the farm household, because it is available locally, there is little or no paperwork and quick service,
(ii) the closeness of the lender helps him understand household needs and judge whether the farmer is credit worthy,
(iii) less risk of default because the lender can easily remind the borrower and use other pressures to make him repay,
(iv) low administrative costs.

The **disadvantages** are:

(i) the borrower may feel under an obligation to the lender and lose his independence,
(ii) there may be a few people able to lend so no choice for the borrower,
(iii) only short-term loans of fairly small amounts are available.

Formal credit sources are institutions, such as banks and co-operative unions. They have standardised lending procedures and are controlled by the Central Bank. Formal credit agencies face special problems in lending to farmers. Administrative costs and risks are high:

(i) Farms are widely scattered and costs of travel to the villages may be high.
(ii) Farmers often want small loans. The cost of administration and supervision of a small loan is the same as for a large one.
(iii) Farming is risky. Crop yields and prices vary a lot. Sometimes farmers may have difficulty in keeping up repayments.
(iv) Farmers lack suitable collateral security. Land, buildings and crops may be pledged or mortgaged. Yet it would be difficult for a credit agency to take away a man's livelihood if he failed to repay his loan.
(v) Many farmers lack knowledge and experience of applying for formal loans.

6.5 Improving Farm Credit

A Government, wishing to improve the supply of farm credit, may set up a special Agricultural Development Bank. Funds are provided by the Government at little or no interest, to provide a cheap banking service for farmers. Branches are opened in rural areas to accept deposits of farmer's savings and to provide farm credit. Development bank agents have special knowledge of agricultural production and marketing. In fact, with government assistance, the Commercial Banks can be encouraged to provide a similar service. This may avoid the cost of setting up a separate agricultural banking system.

Administrative costs and the risks of lending can be reduced in other ways. This allows a better and cheaper service to be provided to farmers.

(i) Lending through Marketing Boards or other produce buyers This works particularly well for short-term loans offered at the start of the cropping season. Repayment is collected when the crop is sold. Rules for repayment should be flexible so that the farmer does not have to sell his crop immediately after harvest. In fact credit may be needed to cover costs of storage. The farmer can then choose to sell when the price is high.

(ii) Group loans Lending to co-operatives or other groups has advantages over lending to individuals. First, a larger sum is lent so administrative costs are a smaller part of the total. Second, farmers within the group can learn from each other how to make use of loans. Third, the group take responsibility for repayment so there is less risk. In fact the system of co-operative unions described in Chapter 3 (Figure 3.3) can be used as a banking system. The central co-operative bank can allocate funds to co-operative unions which in turn provide loans to village societies.

(iii) Using Development Project Agencies The main advantage here is that credit can be linked with other services. For instance new inputs can be made available along with the credit to buy them. In fact the inputs can be provided as credit in kind. At the same time the agricultural extension service may give advice. Extension agents can help farmers draw up plans and credit applications. They can also supervise to make sure the credit is used wisely. All these services may be provided together at Agricultural Service Centres.

7 Agricultural Loans

7.1 The Decision to Borrow

Consumption loans are sometimes needed in an emergency, or for a special occasion. The borrower then finds himself in debt with no extra income to provide for repayment. For this reason many people do not like to borrow unless it is absolutely necessary. Credit which is used to increase production should be viewed as a normal part of good farm management. But its use should be planned in advance. It is no good seeking a loan without first deciding what it will be used for.

In fact the first step is to find a suitable project for investment. This means some agricultural or marketing activity that is likely to be profitable. It may be the establishment of a new crop or livestock enterprise. It may be the used of new methods and equipment. If there are several possible alternative projects, so much the better. They can then be sorted into order and the best one chosen.

The next step then is to work out a budget for each alternative. Methods of preparing budgets were described in Chapter 5. A detailed plan must be worked out for each project and the costs and benefits must be estimated. Cash-flow analysis can be used to estimate the returns. Generally the project that offers the highest return will be chosen. Other points may be taken into account. For example, a project which gives a quick return, may be preferred even if the rate of return is not the highest. Also some projects may be avoided because they are too risky.

When an investment project has been chosen and a budget shows it will earn a worthwhile return, the next step is to decide whether credit is needed. The farmer's own savings may be sufficient. By using his own savings he saves himself the trouble of seeking a loan, and does not have to face interest and repayment costs. Alternatively the funds may be raised by co-operation. A group of farmers may combine their savings to make the investment.

If the farmer or the group does not have sufficient funds to cover the whole investment a loan will be needed. It will only be worth using credit if the expected rate of return is higher than the rate of interest. Then the farmer, or the group, profits from the loan, even though interest is paid. Let us suppose that this is the case, so it is decided to apply for a loan.

7.2 Conditions for Lending

Banks which provide agricultural credit may employ specialist advisors. The agricultural specialists are attached to certain branches of the bank.

They assist farmers applying for agricultural loans and supervise the use of such loans. The bank's primary objectives are to:

(i) provide funds for part-financing the establishment, reactivation, expansion and modernisation of viable agricultural projects (this means projects which earn an acceptable rate of return),

(ii) check, through supervision, that the funds are used for agricultural purposes,

(iii) make sure that the loan and interest charges are repaid promptly,

(iv) monitor progress of the project, assist clients where necessary and offer advice.

Loans may be offered to individual farmers, public or private companies with legal status, partnerships, Governmental and private organisations with legal status and borrowing powers, **or** co-operatives, unions and similar societies.

To qualify for an agricultural loan several conditions should normally be met.

1.An Individual Applicant

(i) He must possess the ability, training and experience needed to manage the project. An absentee farm owner must employ an able manager.

(ii) The applicant must be able to show that the project is viable.

(iii) He must be able to provide the necessary land for the establishment of the project.

(iv) He must be prepared to make a substantial contribution to the cost of the project from his own savings. (The bank will decide how big the borrowers contribution should be.)

(v) In addition the bank may require security for the loan. Special consideration may be given to applications for loans under an Agricultural Credit Guarantee Scheme where it exists.

2. Group Farmers

The same conditions, as listed above, apply to group loans. In addition, the following conditions may be required:

(i) A society or partnership must be legally registered.

(ii) Credit sought by limited liability companies or corporations must be within the borrowing powers of the organization.

(iii) Farming must be one of the functions of the organization and this should be stated in its Memorandum and Articles of Association. This is particularly important for loans guaranteed under an Agricultural Credit Guarantee Scheme.

3. The Project

(i) It must be an agricultural project or a closely allied processing or marketing project. Of course, it should conform with national laws and policies.

(ii) It must be economically viable, and socially desirable.

7.3 Application Procedure

The application procedure may vary in detail from one credit agency to another. The following outline gives a general guide.

Applications should be submitted to the bank at least three months before funds are required. The application should be made to the Branch Manager of the local branch of the bank. The applicant may already have an account at a distant branch. In that case his Branch Manager should introduce him to the manager of the branch nearest the project.

The first approach may be by letter or a personal meeting with the Branch Manager. At some stage an appliation form for credit facilities will have to be completed. This form requests information about the applicant and his family (or the group for group loans), his farm operations, the amount of loan required and its purpose. Three copies of the form must be completed.

Three copies of the Feasibility Report of the project must be submitted with the loan application form, if more than $ 5000 is requested. For smaller loans a Feasibility Report may not be needed. The main purpose a Feasibility Report is to show that the project is viable, and will earn a return high enough to cover the costs of a loan and still show a profit. It should include a full description of the proposed project and cash flow budgets. (See Chapter 4). A farmer is wise to work out these details even if the bank does not require a Feasibility Report.

The applicant may also be asked to produce **balance sheets** for recent years. A balance sheet is based on farm and other business accounts (see Chapter 5) and compares the value of the farmer's capital (his assets) with what he owes (his debts or liabilities). The balance, or net worth, is the difference between assets and liabilities. It is a measure of his wealth or the assets he owns (see Fig 7.1). A balance sheet can equally well be calculated for a group or co-operative, a Limited Liability Company or partnership or any other business organisation. It is a useful guide for business management, not only when seeking a loan.

The balance sheet shows how much the applicant can afford to contribute to the project from his own assets. Only liquid assets such as cash in hand or in a bank account or amounts due from debtors can be used for new investment. Fixed assets are not available for this purpose. Some liquid assets are needed anyway to cover working capital needs. If too much gets tied up in fixed assets leaving insufficient working capital, problems arise. This situation is called **overtrading**.

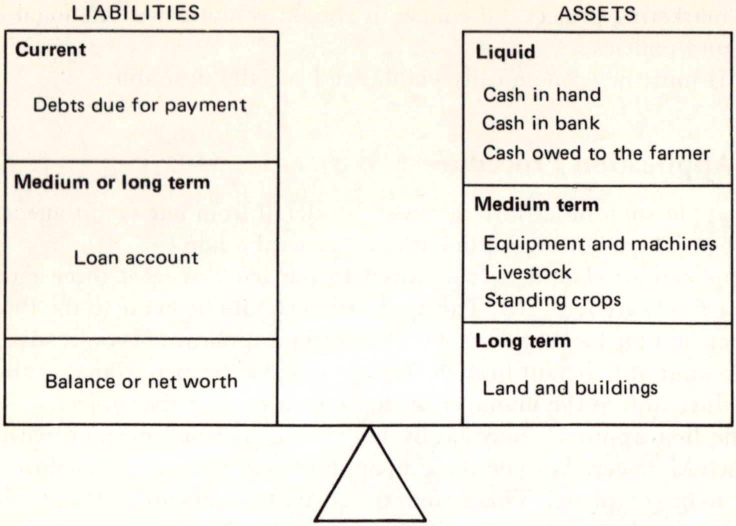

Based on Dalton, G.E. (1982) *Managing Agricultural Systems* Applied Science Publishers, Figure 6.2.

Fig 7.1 The balance sheet

At the same time there are dangers in borrowing too much in relation to net worth. The more that is borrowed the greater are the risks for both borrower and the lender. If crops should fail or disease breaks out, it is harder to repay a big loan than a small one. A useful measure, that can be calculated from the balance sheet, is total debt divided by the net worth. The farmer and the bank manager may want to check that this does not rise too high.

Other documents may be required by the bank. For corporate bodies copies of the Articles and Memorandum of Association may be needed together with Registration Certificates. For government bodies, documents must show the decree or law establishing them and the way in which they are funded.

When all the necessary documents are available the bank will process the application. If the loan is agreed in principle, the bank will need to check on the value and legal status of any security offered. The borrower will need to open a bank account if he does not already have one. Arrangements for repayment will have to be agreed. Finally the farmer is informed where and when to sign the loan agreement and collect the loan.

7.4 Repayment

The disbursement, or payment of the loan should, as far as possible match the farmer's cash needs. Similarly the repayments should be planned for periods when the main returns from the project are expected.

For short-term loans the best time of disbursement is at the beginning of the cropping season when inputs of farm supplies are needed. Repayment may be in a lump-sum after the crop is sold. This may occur at harvest or later if the crop is stored.

For medium and long-term loans the cash flow budgets may show the best times for repayment. Sometimes a lump-sum end repayment may be agreed. Generally, payment by instalments is better. If the project will take a long time to mature, a grace period may be allowed during which no repayments are made. This might be necessary in establishing tree crops for example. After the grace period the loan might then be repaid by a series of equal annual instalments.

Some flexibility of repayment plans is needed. Generally the bank will agree to early repayment, if it is desired by a successful applicant. Sometimes the farmer may wish to delay repayments if he is waiting for prices to rise before selling, or if his project is slow to mature. Clearly the bank cannot allow bad debts to develop, but some flexibility may be possible to help the farmer.

7.5 Summary and Conclusions

This book is concerned with success in farming, which depends on good objectives, decision making, implementation and control. When an agricultural loan is used to finance a new agricultural project all these tasks must be done.

First the farmer must set his objectives. If he wants to maximise profits he will seek loans to expand his business. Then he must make decisions on what to produce, how much to produce and how to produce it. The chapters on basic economic principles and marketing provide a background for making plans. Before deciding to produce something it is important to study the markets and judge whether there is a demand. The law of diminishing returns suggests that inputs must be used in the right balance.

Records and accounts, described in Chapter 4, are very useful for planning purposes. Decisions about what to produce are helpful if there are records of crop and livestock production in the past. Financial accounts can be useful for estimating gross margins or cash flows for planning purposes.

Planning and budgeting, described in Chapter 5, are the basis for better decision making. Before seeking a loan it is essential to draw up a capital budget for the planned project, as described in Section 5 of that Chapter.

Implementation of the planned project may require a loan. The background to agricultural finance and development is discussed in

Chapter 6. This information may be helpful in deciding where to go for a loan and what terms to seek. A good understanding helps good working relationships between credit agent and farmer.

Finally once a loan has been agreed, together with the terms and conditions, the plan is put into action. From this stage on, control becomes important. Overspending on inputs, wastage or loss of output can cause the project to fail. Here again, good records and accounts are important to keep a check on progress. They provide a feedback of information needed for the proper control and management of a project.

It is hoped that this small book provides some useful guidance on all these aspects of good management, and that it will bring you success in farming.

Success in farming begins here – with the practical, basic essentials of agricultural economics and farm management. This book has been written in a plain understandable style for the producer – farmers and farm managers – for advisers, extension staff and for students and teachers whether at senior school, agricultural college or university. Anyone starting up farming as a business will find its advice helpful. The basic economic principles and factors of production are given first, followed by a discussion of the practical aspects of marketing, business analysis, planning, records and accounts. The book then finishes with advice on agricultural finance and how to go about obtaining loans.

Martin Upton is a lecturer in agricultural economics with wide tropical experience. He has written several books on farm management and agricultural economics for students.

ISBN 0-333-38335-4

9 780333 383353

Macmillan Publishers